Diseases of the
Wistar Rat

This book is dedicated to Professor Anthony Dayan with grateful thanks for all his help and encouragement over many years.

Diseases of the Wistar Rat

MARY J. TUCKER

Zeneca Pharmaceuticals

Taylor & Francis
Publishers since 1798

UK Taylor & Francis Limited, 1 Gunpowder Square, London EC4A 3DA
USA Taylor & Francis Inc., 1900 Frost Road, Suite 101, Bristol, PA 19007

British Library Cataloguing in Publication Data

A catalogue record for this book is available from the British Library.

ISBN 0 7484 0521 6

Library of Congress Cataloging in Publication Data are available

Cover design by Jim Wilkie

Typeset in Times 10/12 by M.C.S. Ltd, Salisbury, Wilts

Printed in Great Britain by T. J. International (Padstow) Ltd, Cornwall

Contents

Preface   ix

Acknowledgements   xiii

1 The Alderley Park Rat   1
   1.1   History   1
   1.2   Health status   3
   1.3   Husbandry   4
   1.4   Growth   5
   1.5   Organ weights   7
   1.6   Chemical pathology   9
   1.7   Haematology   9
   1.8   Clinical chemistry   13
   1.9   Urinalyses   14
   1.10 Hormone levels   15
   1.11 Clinical observations   16
   1.12 Mortality   18
   1.13 Histopathology   19
   1.14 References   19

2 The Integumentary System and Mammary Glands   23
   2.1   Skin   23
   2.2   Zymbal's glands   30
   2.3   Preputial glands   30
   2.4   Mammary glands   31
   2.5   References   34

Contents

3 The Musculo-skeletal System   37
   3.1 Muscle   37
   3.2 Joints   40
   3.3 Bone   40
   3.4 References   44

4 The Digestive System   47
   4.1 Oral cavity   47
   4.2 Tongue   51
   4.3 Salivary glands   51
   4.4 Oesophagus   55
   4.5 Stomach   56
   4.6 Intestines   59
   4.7 Liver   62
   4.8 Exocrine pancreas   72
   4.9 References   75

5 The Urinary System   81
   5.1 Kidney   81
   5.2 Ureter   91
   5.3 Urinary bladder   91
   5.4 References   93

6 The Cardiovascular System   97
   6.1 Heart   97
   6.2 Vascular system   105
   6.3 References   107

7 The Respiratory System   111
   7.1 Nasal cavities   111
   7.2 Larynx and trachea   112
   7.3 Lungs   112
   7.4 References   120

8 The Haemopoietic and Lymphatic Systems   123
   8.1 Bone marrow   123
   8.2 Spleen   126
   8.3 Lymph nodes   129
   8.4 Thymus   137
   8.5 References   142

9 The Female Genital System   145
   9.1 Ovaries   145
   9.2 Uterus   154
   9.3 Vagina   158
   9.4 References   159

10   The Male Genital System   163
       10.1   Testes   163
       10.2   Epididymides   171
       10.3   Seminal vesicle   172
       10.4   Prostate   174
       10.5   Coagulating gland   177
       10.6   Preputial glands   177
       10.7   Penis   178
       10.8   References   178

11   The Endocrine System   183
       11.1   Pituitary gland   183
       11.2   Pancreatic Islets of Langerhans   191
       11.3   Adrenal gland   193
       11.4   Thyroid gland   201
       11.5   Parathyroid gland   208
       11.6   References   209

12   The Nervous System   217
       12.1   Brain   217
       12.2   Spinal cord   230
       12.3   Peripheral nerves   232
       12.4   References   234

13   Special Sense Organs and Associated Tissues   237
       13.1   Eyes   237
       13.2   Harderian glands   243
       13.3   Exorbital gland   245
       13.4   Ear   245
       13.5   References   245

       Index   249

Preface

The laboratory rat (*Rattus norvegicus*) has been used extensively during the twentieth century, and many strains have been bred to satisfy the requirements of workers in the different areas of biomedical research. In toxicology there has been much debate as to whether outbred, heterogeneous rats should be used, as the human population is heterogeneous; others have argued that using inbred or hybrid strains is more appropriate as this narrows the statistical confidence limits. The outcome is that laboratories use many different strains and it has been suggested that this has affected the reproducibility of results from one laboratory to another. Toxicologists and research workers are reluctant to change strains for several reasons. An extensive database may have been collected, including reference ranges for many parameters, such as tumour incidences and mortality rates. Published data on these parameters show considerable differences between the same strains in different laboratories, particularly for clinical pathology (Mitruka and Rawnsley, 1981; Ringler and Dabich, 1979). Caging facilities are also geared to specific strains as there is a considerable difference in size between the strains. Toxicologists are aware that the spontaneous patterns of disease vary from strain to strain; the fact that patterns of disease change with time, within a strain, is less well accepted as many laboratories do not breed their own rats, and may use several different commercial breeders. No matter which strain is used, it is essential that all workers with the laboratory rat have a detailed knowledge of the strain they use, including the husbandry, patterns of spontaneous disease and mortality, and it is important that these data are continually monitored. The purpose of this volume is to provide a comprehensive source of knowledge on the spontaneous diseases of a Wistar rat strain which has been bred and used continuously in one laboratory since 1957. It is not intended to be a detailed account of the pathology of the

Table I Numbers of AP rats included in the database

Duration of study	Number of animals in the database[a]	
	Males	Females
1–2 weeks	300	300
1 month	300	300
3 months	500	500
6 months	300	300
12 months	150	150
24 months	2800	2500
Life span (52 months)	192	288
Total	4542	4338

[a] All animals are from control groups dosed with an excipient, by gavage, intravenous administration or subcutaneous depot, or fed an unmedicated diet. All animals were 6–8 weeks old at the start of treatment.

diseases as there are several excellent monographs on this topic, but it will include sufficient pathological description to enable the reader to compare the diseases with published sources. The database of control rats in this monograph includes 4542 males and 4338 females used in various research and toxicology studies between 1960 and 1992. The duration of the studies extends from one week to a life-time study which lasted 52 months; approximately half of the database is from animals in 24 separate oncogenicity studies each of 24 months duration. The number of animals examined at each time point is shown in Table I. Each control group represents only 25 per cent of the total number of animals in each study; the other 75 per cent of animals (approximately 15 000) dosed with the test material support the information in the database but have not generally been included, to avoid any bias. Any data which are not derived from control animals will be specified. In long term studies the major effect of most test compounds on spontaneous disease is to reduce the incidence of common diseases such as tumours and renal disease; this is because doses are selected on the basis of a reduction in body weight, which is known (Tucker, 1979) to reduce the incidence of these diseases. To reduce variation in pathological diagnosis, all of the histopathological examination of the animals included in this database was done by the writer. Comparisons with other strains are included for the more important diseases.

Mary J. Tucker
Macclesfield

References

MITRUKA, B. M. and RAWNSLEY, H. M. (1981) *Clinical, Biochemical and Hematological Reference Values in Normal Experimental Animals and Normal Humans*, New York: Masson Publishing.

RINGLER, D. H. and DABICH, L. (1979) Hematology and clinical chemistry, in BAKER, J. H., LINDSEY, J. R. and WEISBROTH, S. H. (Eds), *The Laboratory Rat* Vol. 1 , pp. 105–21, New York: Academic Press.

TUCKER, M. J. (1979) The effect of long term food restriction on tumours in rodents, *International Journal of Cancer*, **23**, 803–7.

Acknowledgements

I thank all of the staff of ICI Pharmaceuticals Division who worked in safety evaluation, first in the Toxicology Unit of the Research Department and then in the Safety of Medicines Department and the staff of the Library. My thanks also to Margaret Bowles for her help with preparation of the manuscript.

1

The Alderley Park Rat

1.1 History

The purpose of this chapter is to provide the reader with basic information on the origins of the strain, the husbandry and standard laboratory data since even these parameters may differ from one Wistar rat colony to another. The Alderley Park rat is an outbred, albino, specific pathogen free (SPF) Wistar-derived rat which has been bred and used continuously at the Zeneca/ICI laboratories at Alderley Park, Macclesfield since 1957. The rat is designated the Alpk:AP$_f$ SD (Wistar derived) (AP rat) strain, short form AP rat, and it has been a closed colony since 1957. The albino Wistar rats were first established at the Wistar Institute in 1920; from that colony another was established at Porton laboratories and the AP strain was derived from a breeding nucleus supplied from the Porton colony in 1942. At first the strain was maintained in conventional (non-SPF) conditions at ICI Blackley, Manchester until the laboratories of ICI Pharmaceuticals Division were opened, at Alderley Park, in 1957. The SPF breeding unit for rodents was on the same site, but separate from the experimental laboratories where the animals were used. Details of the building and maintenance of the SPF breeding unit were described by Davey (1959). The AP rat colony was established with a small number of litters which were Caesarian-derived from the conventional colony. The newborn rats were dipped in antiseptic, and passed into the SPF unit where they were hand reared. Initially the rats were bred on a harem basis (one male and four females), then on a pair basis.The females have six litters, all force weaned on day 21, and after the sixth litter the breeding pairs are discarded aged approximately 14 months. The rats are bred according to a statistical plan which avoids brother/sister and parent/offspring matings. Although the AP rat is designated an outbred strain, DNA fingerprinting suggests that the genetic variation between

Diseases of the Wistar Rat

Table 1.1 Reproductive indices in the AP rat in the period 1990–95

Total number of dams	326
Mean litter size	12.21
Mean foetal weight	3.7 g
Mean placental weight	0.51 g
Sex ratio (proportion male)	0.51
Pre-implantation loss	3.4
Post-implantation loss	1.43

Table 1.2 Spontaneous foetal abnormalities in the AP rat

Abnormality	Mean incidence in 3774 foetuses[a] (%)
Haematoma	0.57
Cleft palate	0.05
Oedema	0.19
Agnathia	0.05
Anal atresia	0.05
Anophthalmia	0.05
Microstomia	0.05
Synotia	0.05
Kinked ureter	1.48
Dilated ureter	2.14
Renal pelvic cavitation	0.19
Extra liver lobe	1.76
Dilated cerebral ventricles	0.14
Elongated thymus	0.19
Great vessel anomalies	0.05
Situs inversus	0.05
Mis-positioned umbilical artery	11.71
Mis-positioned azygous vein	0.05
Absent mandible	0.05
Extraneous dot palate	0.43
Irregular ridging palate	0.29
Coiled tail	0.05
Craniocele	0.05
Cranioschisis	0.05

[a] Incidence (mean of values from 18 reproductive studies) of spontaneous foetal abnormalities in 3774 foetuses of AP rats examined between 1990 and 1995. Total incidence of all abnormalities is 11%.

individuals is little different from that of an inbred strain. Reproductive indices for the strain in the period 1990–95 are shown in Table 1.1 and the incidence of spontaneous foetal abnormalities in Table 1.2. The colony has been re-derived on several occasions over the 40-year period due to infections, both viral and bacterial. This was done by fostering Caesarian-derived AP rats onto Charles River gnotobiotic animals, maintained in isolators, until sufficient numbers were available to re-stock the breeding areas. This was also the method used to stock a second, separate, SPF breeding unit built at the end of the 1970s.

The breeding unit maintains several other strains of rats, mice and guinea pigs and supplies animals for both research and toxicology requirements. Animals used in the toxicology studies are kept in separate facilities from the animals used for research studies.

1.2 Health Status

The strain is monitored every three months for genetic drift using the mandibular technique of Festing (1974). In the first few years after the colony was established it was only monitored for the virus of chronic respiratory disease, from which it remained free. In the second decade the colony was infected with *Pasteurella pneumotropica* which produced a high mortality from bronchopneumonia, particularly in weanling rats, and the breeding animals were treated with antibiotics for several years. At the end of the 1960s the colony was infected with Sendai virus and there was a high mortality in weanling rats; further outbreaks occurred sporadically until the colony was re-derived. From that time the re-derived colony has been vaccinated with inactivated Sendai virus. The AP rat is monitored at three-monthly intervals for the following microbes: bacteria – *Staphylococcus pyogenes*, *Streptococcus pneumoniae*, *Klebsiella pneumoniae*, *Bordetella bronchisepta*, *Neisseria* species, *Pasteurella multicida*, *Pasteurella pneumotropica*, *Mycoplasma* species, *Streptobacillus moniliformis*, *Pseudomonas aeruginosa*, *Salmonella* species; viruses – Sendai, minute virus of mice, Polyoma, Reovirus III, Toolan's HI, Kilham adenovirus, mouse hepatitis virus, rat corona virus. Faecal samples are examined for protozoans and helminths. In toxicology studies groups of untreated sentinel animals are kept in the same room throughout the study; this is to provide a source of serum for monitoring microbes, if unexpected ill-health occurs, without disturbing the animals of the toxicology study.

Animals for toxicology studies are received from the Animal Breeding Unit when they are approximately 4 weeks old. They have an acclimatisation period of 7 to 9 days during which they are observed for signs of ill health. At the end of the acclimatisation period the required numbers of animals are randomly selected and the remainder are removed from the animal room before the dosing period commences.

3

1.3 Husbandry

Husbandry has changed to some extent, during the existence of the colony, as different standards and government regulations have been introduced. The SPF breeding unit has always had barriered production areas where rats of different strains are bred, but the experimental animals, which are also barrier maintained, are not mixed with any other animals. Each study is kept in a single room and no other animals are introduced during the period of experimentation. This has meant that no serious infection has occurred in the experimental animals, except for those infections which were acquired in the breeding unit and transferred to the experimental rooms. Over the years there have been significant changes in regard to caging. This has always been in stainless steel cages, with mesh floors, suspended in racks over paper-lined steel trays; the papers are changed daily and the cages changed and sterilised every second week. Cage sizes have increased and gang housing has decreased, from five in cages sized $45 \times 28 \times 20$ cm, to 3/cage for males and 4/cage for females in cages of $57 \times 35 \times 20$ cm. At first animals were housed in rooms exposed to normal daylight and artificial lighting, but new experimental procedures, which were introduced in 1970, established a 12/12 hour artificial lighting/dark cycle. Environmental conditions are monitored daily and are aimed at maintaining an ambient temperature of $20 \pm 2\,°C$ and humidity at 55 ± 10 per cent. Water is supplied by automatic systems from the site drinking water

Table 1.3   R&M No 1 (modified) diet

Dietary constituents
Wheat
Wheat feed
Barley
Soya bean meal extract
Whey powder
Soya oil
Vitamin/mineral/amino acid premix

Major nutritional components	Calculated analysis (%)
Crude protein	14.7
Crude oil	2.6
Crude fibre	5.3
Ash	5.9
Calcium	0.55
Phosphorus	0.50
Magnesium	0.21

Table 1.4 Dietary contaminants specification

Chemical	Maximum permitted concentration (ppm)	Microbe	Maximum permitted level
Arsenic	1.0	Total organisms	$1 \times 10^2/g$
Cadmium	0.2	Mesophilic spores	$1 \times 10^2/g$
Lead	3.0	*Salmonella* sp.	None/50 g
Mercury	0.1	Faecal *E. coli*	None/50 g
Selenium	0.5	(type 1)	
DDT (total)	0.1	Coliforms	None/50 g
Dieldrin	0.02	Fungal units	None/10 g
Heptachlor	0.01	Antibiotic activity	None/g
Lindane	0.1		
PCBs (total)	0.05		
Fluorine	40.0		
Nitrite	5.0		
Nitrate	100.0		
Aflatoxins (total)	0.001		
Malathion	0.5		

supply and diets are pelleted on site from powdered material provided by different commercial sources. From 1957 to 1975 AP rats were fed Powder 'O' diet, a diet made to our own specifications and compounded by a local miller (Oakes, Congleton, UK). The formulation has been published (Tucker, 1979). From 1975 a commercial diet (PCD diet, BP Nutrition, UK) was used. In the last decade irradiated R&M No 1 (modified) diet (Table 1.3), manufactured by Special Diet Services (Witham, UK) has been used in all toxicology studies. All of these diets are 'natural', cereal based diets, with the major difference that R&M No 1 diet is a maintenance diet with a protein level of 14.7 per cent compared with the high protein level of approximately 25 per cent in Powder 'O' and PCD diets; in addition the fish meal in Powder 'O' was replaced by soya in the other two diets. All batches are required to be tested and to comply with a contaminants specification (Table 1.4). Any batch not complying with the specification is discarded.

1.4 Growth

Growth curves show that approximately 75 per cent of growth occurs in the first 16 weeks of a study, i.e. up to the age of 22 to 24 weeks of age. The remaining 25 per cent of growth is much slower until 80 weeks, when the onset of spontaneous diseases affects body weights. This growth pattern is similar to that described for the Charles River Sprague-Dawley (SD) rat by Lang and White (1992), although their data were derived from a variety of different laboratories. It has also been shown, by Klinger *et al.* (1996), that there is variation in growth of three strains of SD rats which were fed the same diet. The maximum size of the SD and AP rats is similar but much greater than that of the F3444 rats (Cameron *et al.*, 1985; Lang and White, 1992). During the 40-year period the AP rat has been used, there has been a considerable increase in body weight, particularly in males, as shown in Figure 1. This compares growth curves in control animals (100/sex) from two-year studies, one completed in 1971 the other in 1992. The maximum mean body weight of the male animals in the 1971 study was 550 g and in the 1992 study was 800 g, and in females 360 g in 1971 and 510 g in 1992. There are probably several factors which have contributed to this increase in size. In breeding colonies there is always pressure to produce animals of the required weight as quickly as possible. When selecting animals for breeding, the larger animals from a litter are usually chosen and breeding animals and their offspring are also fed diets designed to promote rapid growth. Over several generations these factors, together with overnutrition and lack of exercise, have contributed to the increase in the size of the whole colony of AP rats.

The occasional outbreak of infectious disease in the first two decades must have contributed to the lower body weight, thus the elimination of significant infection is another factor which has promoted body growth. The consequence has been that AP rats more than 12 months old are obese, particularly males, with extensive deposition of fat in abdominal fat pads and subcutaneous areas. The introduction of a low protein maintenance diet for all toxicological studies,

Comparison of body growth curves in 1969 and 1992

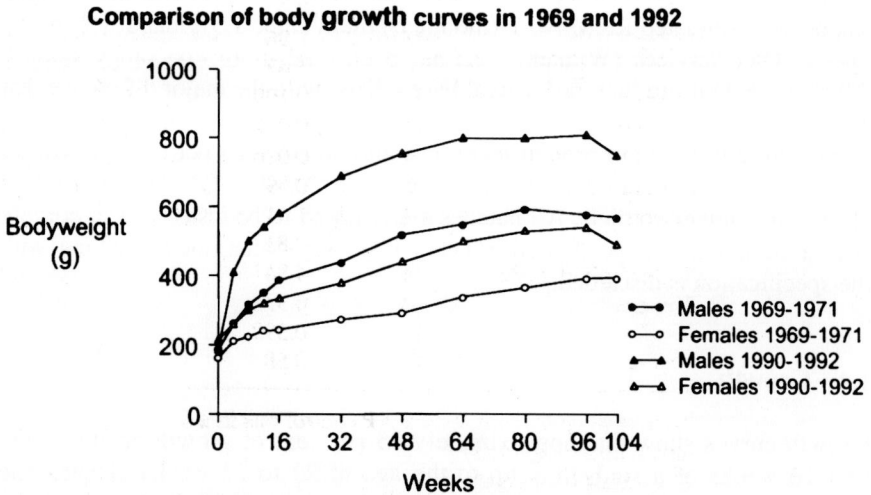

Figure 1 Comparison of body growth curves in the AP rat.

in the last decade, is aimed, primarily, at reducing the mortality in males from renal disease, but it is hoped that it will also reduce growth in the experimental animals.

1.5 Organ Weights

It is not possible to give a reference range of organ weights which could be applied to all Wistar rat strains since many organs grow in relation to body weight and will, therefore, be dependant on several factors, such as diet, which can alter body weights significantly. The absolute weights of most organs do not change substantially once the animals have passed the 16-week period of rapid growth. An example of absolute organ weights from animals

Table 1.5 Organ weights in control animals

Organ[a]	Sex[b]	Mean weight[c] (g)
Adrenal glands	M	0.066
	F	0.099
Brain	M	2.16
	F	2.00
Heart	M	1.80
	F	1.18
Kidneys	M	3.71
	F	2.50
Liver	M	22.0
	F	12.3
Lungs	M	1.96
	F	1.47
Ovaries	F	0.109
Pituitary gland	M	0.012
	F	0.016
Prostate	M	0.59
Spleen	M	1.11
	F	0.81
Testes/epididymides	M	4.84
Thymus	M	0.338
	F	0.399
Uterus	F	0.58

[a] Organ weights from 20/sex AP control rats in a 6 month toxicology study.
[b] M = Male F = Female
[c] Organ trimmed free of extraneous tissue and weighed before fixation.

Table 1.6 Comparison of relative organ weights

| Organ[a] | Sex[b] | Organ weight as % of body weight[c] | | |
		1 month	6 months	12 months
Adrenal glands	M	0.017	0.011	0.008
	F	0.035	0.030	0.021
Brain	M	0.56	0.36	0.30
	F	0.78	0.62	0.51
Heart	M	0.36	0.29	0.23
	F	0.40	0.36	0.29
Kidneys	M	0.73	0.61	0.52
	F	0.76	0.71	0.64
Liver	M	5.0	3.7	3.0
	F	4.5	3.8	3.4
Lungs	M	0.41	0.32	0.27
	F	0.50	0.45	0.39
Ovaries	F	0.045	0.033	0.027
Pituitary[c]	M	2.7	1.9	3.6
	F	4.8	5.2	4.9
Prostate	M	0.097	0.098	0.089
Spleen	M	0.25	0.19	0.16
	F	0.27	0.25	0.21
Testes/epididymides	M	1.16	0.79	0.72
Thymus	M	0.15	0.059	NR
	F	0.21	0.018	NR
Uterus	F	0.12	0.17	0.15

[a] Comparison of relative organ weights (organ as % body weight) in control AP rats in a 1 month study (10/sex), a 6 month study (20/sex) and a 12 month study (25/sex).
[b] M = male F = female.
[c] Pituitary + % × 1000.
NR = not recorded; the thymus gland is not weighed in studies longer than 6 months duration.

in a 6 month study is given in Table 1.5. The only organ weight which is significantly greater in females, compared with males, is the adrenal gland weight, and this gender difference has also been reported for other strains of rat such as the Sprague-Dawley (Yarrington and Johnston, 1994). A comparison of relative organ weights in a 1, 6 and 12 month study is given in Table 1.6 and demonstrates the decrease in relative organ weight, with time, due to the continual growth of the animals to 12 months. The pituitary weights are an exception and show no clear decline in relative weight, which is probably related to individual differences in hormonal activity and the onset of proliferative disorders.

1.6 Chemical Pathology

The usual methods of blood sampling for clinical chemistry in the AP rat are to take samples from the tail veins during life, and from the vena cava at necropsy. Other methods which have been infrequently used include cardiac puncture and orbital-sinus bleeding, but this latter method is not used when histopathology of the eyes and adnexae are required. Values for haematology and clinical chemistry parameters have shown some variation with time, and methods of analysis have altered significantly with the periodic introduction of new analytical machines. Reference ranges have been updated continually with the most recent update for the years 1991–94. The analyses are from toxicology studies of 1, 3, 6 and 12 months duration and have been combined at time points which do not show any statistically significant differences. Data for older animals are not available as clinical chemistry analyses were not done routinely in carcinogenicity studies in the ICI laboratories because of the confounding effects of age-associated diseases.

1.7 Haematology

Reference ranges for haematology analyses in AP rats are shown in Tables 1.7 and 1.8 (coagulation factors). Haematology data for many strains have been published (Payne *et al.*, 1976; Lewi and Marsboom, 1981; Weil, 1982; Leonard and Ruben, 1986), and the reference ranges for the AP rat are in general agreement. This confirms the observations of Bailly and Duprat (1990), who noted no significant differences in their review of haematological parameters in several strains of rat. Minor variations between strains are most likely due to differences in sampling, anaesthesia, analysis and nutrition (Upton and Morgan, 1975; Archer and Riley, 1981; Yamanuchi *et al.*, 1981; Suber and Kodell, 1985). In the AP rat the only important changes with time are in haemoglobin levels and total red cell counts – which increase, with values slightly higher in males. Total white cell counts are also higher in males but they do not change with time. Differential counts show a significant change as the animals age, with lymphocytes decreasing and neutrophils increasing. Reference ranges for coagulation factors (Table 1.8) are only available for animals from 1 and 3 month studies. They are slightly lower in females.

Femoral bone marrow smears are taken from all AP rats in toxicology studies and, in agreement with data published for other strains, show few non-neoplastic abnormalities. Atrophy of the bone marrow increases with age and there is replacement of cellular marrow by adipose tissue, or fibrosis in some disease conditions. Spontaneous atrophy is more common in females, and at 2 years all animals will show some degree, although it is only severe in about 5 per cent. Neoplastic diseases vary greatly with strain. In the AP rat leukaemia is very rare with acute myeloblastic leukaemia the most common type, followed by acute lymphoblastic leukaemia and the least common type,

9

Table 1.7 Reference ranges for haematology parameters in the AP rat

Parameter[a]	Unit	Sex[b]	Time (months)	Lower 0.5%[c]	Lower 2.5%	Median	Upper 2.5%	Upper 0.5%
Haemoglobin	g/dl	M	1,3	11.5	12.1	14.9	15.8	16.2
			6,12	13.4	13.9	15.2	16.0	16.1
		F	1,3	11.3	11.0	14.4	15.8	16.2
			6,12	13.4	13.9	15.2	16.0	16.1
Red blood cells	10^{12}/l	M	1,3	5.7	6.0	8.3	9.5	9.9
			6,12	6.2	8.0	8.9	9.8	10.1
		F	1,3	5.5	5.9	7.7	8.6	9.2
			6,12	7.2	7.3	8.0	8.6	8.6
Red cell distribution width	%	M	1,3	10.5	10.7	12.5	16.2	18.6
			6,12	11.6	11.8	13.4	15.7	16.5
		F	1,3	9.8	10.4	11.6	15.7	17.6
			6,12	10.8	10.9	12.0	15.1	17.3
Haematocrit	l/l	M	1,3	0.36	0.38	0.44	0.47	0.49
			6,12	0.39	0.40	0.44	0.48	0.50
		F	1,3	0.34	0.36	0.41	0.46	0.47
			6,12	0.36	0.38	0.42	0.46	0.47
Mean cell volume	fl	M	1,3	46.3	47.1	52.8	63.7	66.6
			6,12	45.1	46.1	49.4	53.1	53.5
		F	1,3	49.4	50.4	54.2	61.9	69.6
			6,12	48.4	49.7	52.8	56.9	58.2
Mean cell haemoglobin	pg	M	1,3	16.0	16.2	18.2	20.1	20.7
			6,12	15.9	16.0	17.0	18.1	18.7
		F	1,3	17.1	17.2	18.8	20.4	21.7
			6,12	16.9	17.3	18.4	19.6	20.4
Mean cell haemoglobin concentration	g/dl	M	1,3	30.1	31.6	34.0	36.4	37.9
			6,12	32.2	32.5	34.3	36.6	37.1
		F	1,3	30.0	32.0	34.7	36.7	37.6
			6,12	31.0	33.0	35.0	36.7	37.8
Platelet count	10^9/l	M	1,3	474	625	875	1148	1220
			6,12	452	614	871	1148	1226
		F	1,3	365	601	866	1068	1151
			6,12	531	607	831	1045	1154
White cell count	10^9/l	M	1,3	4.4	5.4	8.0	12.4	14.1
			6,12	4.8	5.0	7.9	13.3	14.6
		F	1,3	2.1	3.3	5.4	8.6	9.4
			6,12	2.2	2.6	4.5	7.7	8.1
Neutrophils	10^9/l	M	1,3	0.752	0.919	1.708	3.398	4.992
			6,12	0.890	1.003	2.229	5.241	7.661
		F	1,3	0.524	0.661	1.308	2.714	3.363
			6,12	0.484	0.572	1.217	3.426	5.429
Lymphocytes	10^9/l	M	1,3	3.114	3.722	5.797	9.039	10.082
			6,12	0.924	3.058	4.810	7.980	8.525
		F	1,3	1.797	2.223	3.753	5.901	7.137
			6,12	1.280	1.443	2.722	4.590	5.107

Table 1.7 *(continued)*

Parameter[a]	Unit	Sex[b]	Time (months)	Lower 0.5%[c]	Lower 2.5%	Median	Upper 2.5%	Upper 0.5%
Monocytes	$10^9/l$	M	1,3	0.040	0.091	0.211	0.438	0.578
			6,12	0.098	0.110	0.243	0.713	0.949
		F	1,3	0.040	0.061	0.125	0.304	0.325
			6,12	0.032	0.050	0.126	0.403	0.737
Eosinophils	$10^9/l$	M	1,3	0.068	0.092	0.191	0.501	0.525
			6,12	0.080	0.143	0.279	0.546	0.717
		F	1,3	0.049	0.063	0.134	0.303	0.429
			6,12	0.059	0.067	0.145	0.354	0.397
Basophils	$10^9/l$	M	1,3	0.000	0.000	0.009	0.025	0.037
			6,12	0.000	0.000	0.004	0.008	0.010
		F	1,3	0.000	0.000	0.005	0.032	0.035
			6,12	0.000	0.000	0.004	0.008	0.010

[a] Haematology parameters derived from control AP rats, 185/sex pooled from 1 and 3 month toxicity studies and 93/sex pooled from 6 and 12 month studies completed between 1990 and 1995.
[b] M = Male F = female.
[c] The upper and lower 2.5% values represent the normal reference range, and the upper and lower 0.5% represent extreme values observed in this time period.

Table 1.8 Coagulation factors in the AP rat

Parameter[a]	Unit	Sex[b]	Lower 0.5%[c]	Upper 2.5%	Median	Lower 0.5%	Upper 2.5%
Prothrombin time	s	M	9.1	9.2	14.7	19.1	20.7
		F	11.9	12.0	14.1	16.2	16.8
Partial thromboplastin time with kaolin	s	M	14.3	19.2	28.8	48.4	51.8
		F	15.2	15.3	24.9	31.7	34.2

[a] Prothrombin time and partial thromboplastin time with kaolin derived from 48 male and female control AP rats used in 1 and 3 month toxicity tests completed between 1990 and 1994.
[b] M = Male F = Female.
[c] The upper and lower 2.5% values represent the normal reference range, and the upper and lower 0.5% represent the extreme values observed in this time period.

monocytic (large granular lymphocyte) leukaemia. The earliest age at which leukaemia was observed was in an animal of 5 months. Loss of weight and body condition precedes death, but the decline into a moribund state is rapid. High peripheral white cell counts and widespread marrow replacement by leukaemic cells occurs in the myeloid leukaemias. Leukaemia will be discussed in greater detail in Chapter 8 (The Haemopoietic and Lymphatic Systems).

11

Table 1.9 Reference ranges for clinical chemistry parameters in the AP rat

Parameter[a]	Unit	Sex[b]	Time (months)	Lower 0.5%[c]	Lower 2.5%	Median	Upper 2.5%	Upper 0.5%
Glucose	mmol/l	M	1	5.7	6.4	8.3	14.4	16.6
		M	3,6,12	5.0	5.4	7.2	9.3	12.2
		F	1	5.5	6.3	8.4	16.0	17.5
		F	3,6,12	5.6	5.7	7.6	9.5	10.2
Urea	mmol/l	M	1	3.9	4.1	5.9	7.3	7.5
		M	3,6,12	4.5	5.0	6.3	7.9	10.0
		F	1	3.7	3.8	6.9	10.1	11.5
		F	3,6,12	5.0	5.6	7.1	10.3	12.2
Total protein	g/l	M	1	48	48	67	73	73
		M	3,6,12	63	66	73	80	81
		F	1	47	47	64	72	73
		F	3,6,12	62	64	72	83	85
Albumin	g/l	M	1	30	31	35	39	40
		M	3,6,12	30	32	35	39	40
		F	1	30	31	35	39	40
		F	3,6,12	32	33	38	44	45
Creatinine	µmol/l	M	1	40	41	53	58	58
		M	3,6,12	47	52	60	70	77
		F	1	39	42	55	64	65
		F	3,6,12	52	55	62	73	86
Alkaline phosphatase	IU/l	M	1	66	249	378	716	733
		M	3	156	181	241	347	437
		M	6,12	140	150	238	470	503
		F	1	119	125	221	606	624
		F	3	77	75	159	257	269
		F	6,12	62	69	119	206	234
Alanine amino-transferase	IU/l	M	1,3	30	33	48	80	91
		M	6,12	27	33	59	120	160
		F	1,3	18	20	42	72	177
		F	6,12	22	28	57	163	238
Aspartate amino-transferase	IU/l	M	1,3	56	59	76	111	138
		M	6,12	59	61	101	236	359
		F	1,3	60	63	79	131	180
		F	6,12	60	63	110	302	592
Cholesterol	mmol/l	M	1,3	1.8	1.8	2.6	3.4	3.6
		M	6,12	2.0	2.1	3.5	6.6	9.6
		F	1,3	1.1	1.5	2.3	3.5	4.1
		F	6,12	1.5	1.7	2.9	4.7	5.7
Triglycerides	mmol/l	M	1	0.49	0.60	1.68	2.49	2.59
		M	3,6,12	0.70	0.91	1.74	3.16	5.12
		F	1	0.26	0.28	0.90	1.45	1.51
		F	3,6,12	0.40	0.60	1.44	3.76	5.92
Sodium	mmol/l	M	1	140	140	144	150	150
		M	3,6,12	139	140	144	149	154
		F	1	136	137	143	150	150
		F	3,6,12	138	138	144	149	154

Table 1.9 *(continued)*

Parameter[a]	Unit	Sex[b]	Time (months)	Lower 0.5%[c]	Lower 2.5%	Median	Upper 2.5%	Upper 0.5%
Potassium	mmol/l	M	1	3.8	3.9	4.7	7.2	7.6
		M	3,6,12	4.0	4.2	4.8	5.7	6.0
		F	1	3.3	3.4	4.6	7.8	9.5
		F	3,6,12	3.5	3.8	4.3	5.5	6.3
Chloride	mmol/l	M	1	93	95	100	108	108
		M	3,6,12	97	98	102	107	109
		F	1	95	96	101	112	112
		F	3,6,12	94	96	103	111	118
Total calcium	mmol/l	M	1	0.52	1.06	2.84	6.64	8.93
		M	3,6,12	2.42	2.50	2.79	3.01	3.06
		F	1	0.81	1.32	2.75	6.39	9.36
		F	3,6,12	2.46	2.51	2.75	3.00	3.08
Inorganic phosphorus	mmol/l	M	1	0.21	1.26	2.39	50.6	100.6
		M	3	1.63	1.65	2.03	2.39	2.49
		M	6,12	1.23	1.28	1.62	2.01	2.11
		F	1	1.11	1.32	2.11	42.8	50.9
		F	3	0.82	1.08	1.56	2.16	2.24
		F	6,12	0.75	0.99	1.35	1.88	2.27
Total bilirubin	µmol/l	M	1	0	0	1	2	2
		M	3,6,12	0	1	2	4	7
		F	1	0	0	1	2	2
		F	3,6,12	0	0	1	3	4

[a] Clinical chemistry parameters derived from control AP rats, 75 male and 78 female from 1 month studies, 80 male and 79 female rats from 3 month studies and 97/sex from 6 and 12 month studies combined. Where no difference was found between 3, 6 and 12 month parameters these were combined.
[b] M = Male F = Female.
[c] The upper and lower 2.5% values represent the normal reference range, and the 0.5% values represent the extreme values observed in this time period.

1.8 Clinical Chemistry

Clinical chemical reference ranges are shown in Table 1.9. Many parameters show little change with time but others do vary. Young rats have a high alkaline phosphatase activity (ALP), which decreases with time, and it is also higher in males than females. The plateau for activity is present by 3 months in males, a little later in females. This has been observed in other Wistar rat colonies (Lewi and Marsboom, 1981), and also in other strains such as SD rats (Charles River Inc., 1989). Minor differences in ALP activity will always be present between strains, as factors such as fasting will decrease activity and high dietary lipid levels will increase activity (Masden and Tuba, 1952; Young *et al.*, 1981).

Total protein levels do show an increase with time, possibly a reflection of the rising incidence of renal disease, although there is no difference between the sexes up to 12 months, whereas renal disease is more common in male AP rats. In male F344 rats a decrease in protein levels has been reported with time (Coleman *et al.*, 1977), while Loeb and Quimby (1989) reported no changes in SD rats up to 24 months and other data on SD rats demonstrate a progressive increase in total protein levels (Charles River Inc., 1989). Albumin levels in AP rats are similar in the sexes and show no changes between 1 and 12 months. In other Wistar colonies, and other rat strains, differences in albumin levels are similar to those which occur with total protein levels: F344 rats showing a decrease with time, and SD rats showing a progressive increase (Coleman *et al.*, 1977; Charles River Inc., 1989), while Lewi and Marsboom (1981) showed a decline in albumin levels in Wistar rats up to 20 months.

Urea levels are similar in the sexes and show no increase up to 12 months of age, which has also been reported for SD rats (Charles River Inc., 1989). Lewi and Marsboom (1981) reported an increase in Wistar rats and Zager and Alpers (1989) a slight increase in F344 rats with time.

Cholesterol and triglyceride levels in AP rats show no changes with time, and conflicting results have been published for other strains. Coleman *et al.* (1977) reported an increase in F344 rats with time, while an increase in SD rats between 1 and 29 months has been reported by Loeb and Quimby (1989); in the report by Charles River Inc. (1989) no change was seen up to 12 months, but thereafter there was a rapid increase.

Serum glucose levels are affected by fasting, even for short periods, and this makes comparisons difficult, but most reports indicate no significant change between the sexes or over time. The following parameters show no major changes in mean values with time in AP rats or any other strains: creatinine, alanine aminotransferase, aspartate aminotransferase, sodium, potassium, calcium, inorganic phosphorus and total bilirubin. Individual levels of all parameters may show significant changes in disease.

1.9 Urinalyses

Reference ranges for quantitative urine analyses are shown in Table 1.10. Urinary sodium, potassium and creatinine all increase with time in both sexes. Non-quantitative assessment of protein has shown a proteinurea at all times, more common in males, and increasing with age. The gender difference is due to the large component of α_{2u}-globulin in the urine, a protein only present in male rats. It is reported that this component decreases with age and albumin becomes the predominant protein in the urine of both sexes (Alt *et al.*, 1980; Horbach *et al.*, 1983). This proteinurea can be related to increasing renal damage in the AP rat, and the changes in electrolytes may also be related to impairment of renal function.

Table 1.10 Urine parameters in AP rats

Parameter[a]	Unit	Sex[b]	Time point (months)	Lower 0.5%[c]	Upper 2.5%	Median	Upper 2.5%	Lower 0.5%
Volume	ml	M	1	2.0	3.0	16.5	34.0	54.0
			3,6,12	4.0	4.0	10.0	27.0	35.0
		F	1	1.0	2.0	11.5	27.0	34.0
			3,6,12	3.0	3.0	7.0	21.0	23.0
Specific gravity[d]		M	1	1.004	1.007	1.016	1.032	1.064
			3,6,12	1.006	1.011	1.034	1.064	1.072
		F	1	1.007	1.008	1.017	1.046	1.080
			3,6,12	1.010	1.011	1.035	1.076	1.078
Sodium	mmol/l	M	1	7.8	13.4	39.9	98.8	150.3
			3,6,12	NA	NA	47.2	124.0	127.2
		F	1	8.7	10.1	29.2	100.2	155.2
			3,6,12	NA	NA	47.2	124.0	127.2
Potassium	mmol/l	M	1	11.2	26.6	83.1	192.9	303.6
			3,6,12	18.1	32.1	156.8	330.0	377.2
		F	1	24.5	26.0	70.6	247.4	407.6
			3,6,12	25.2	25.9	136.4	344.8	381.6
Creatinine	μmol/l	M	1	639	1457	3230	7120	15681
			3,6,12	1426	4057	9748	20352	21285
		F	1	1270	1353	3001	9550	13101
			3,6,12	1864	1935	7622	16908	21041

[a] 90 males and 90 females in 1 month toxicity studies, and 77 males and 69 females from 3, 6 and 12 month studies completed between 1990 and 1994.
[b] M = Male F = female.
[c] The upper and lower 2.5% values represent the normal reference range, and the upper and lower 0.5% represent the extreme values observed in this time period.
[d] Dipstick analysis.
NA = Not available.

1.10 Hormone Levels

Reference ranges for levels of hormones in the AP rat are not yet available but limited data are given in Table 1.11. As would be expected, levels of the gonadotrophic hormones reflect the functional state of the reproductive organs, with luteinising hormone and follicular stimulating hormone decreasing with age, and this has also been reported for F344 rats (Parkening *et al.*, 1983). Values for prolactin and thyroid hormones are only available from rats in 1 month studies. Prolactin levels are four times higher in females than males, while levels of triiodothyronine (T_3) and thyroid stimulating hormone are similar in the sexes, but thyroxine (T_4) is higher in males.

Table 1.11 Various hormone levels in AP rats (1 and 6 month studies)

Parameter	Unit	Number of animals	Sex[a]	Time point (weeks)	Median value
Luteinising hormone	nmol/l	10	M	13	1.86
				25	1.42
				33	1.44
		10	F	13	2.00
				25	1.36
				33	1.50
Follicular stimulating hormone	nmol/l	10	M	13	29.95
				25	62.00
				33	18.00
		10	F	13	11.60
				25	30.95
				33	13.30
Prolactin	ng/ml	5	M	4	27.6
		5	F	4	7.5
Thyroid stimulating hormone	ng/ml	5	M	4	6.6
		5	F	4	6.1
Thyroxine	mol/l	5	M	4	68.7
		5	F	4	38.8
Triiodothyronine	nmol/l	5	M	4	1.87
		5	F	4	1.51

[a] M = Male F = Female.

1.11 Clinical Observations

Clinical signs in rats may be non-specific and indicative only of general ill health, others may be associated with specific diseases. The example of typical clinical signs is taken from a large group of control animals (300/sex) in a 2 year study which was set up to monitor the pathological status of the strain. It was completed in 1984 but is typical of the pattern of observations seen in the AP rat. Table 1.12 lists the clinical signs which lead to the death or unscheduled sacrifice of the animals. Table 1.13 lists the incidence of the less important signs seen in the animals which survived to termination of the study at 2 years.

Firstly, Table 1.12 shows that the most frequent of the important clinical observations was severe weight loss which is assessed as more than a 10 per cent loss of body weight in one week (the animals were weighed weekly). It is a non-specific sign of ill health, but in this study was related to the terminal stages of severe renal disease in males or a large pituitary tumour in females. Severe urinogenital staining was only related to renal disease in a few males, in the majority it was a non-specific sign of general ill health and frequently accompanied the clinical sign of general lethargy. The domed head was caused

16

Table 1.12 Incidence of important clinical observations

Clinical observation	% Incidence[a]	
	Male	Female
General lethargy	8	19
Severe weight loss	1	8
Severe urinogenital staining	12	13
Masses	6	7
Pale extremities	9	5
Excessive salivation	6	9
Malocclusion	1	0.3
Paraplegia	4	1
Laboured respiration	4	3
Domed head	1	0
Ataxia	0.3	4
Abnormal aggression	3	0

[a] Incidence of clinical observations leading to death or unscheduled necropsy in a 2 year study of 300/sex control AP rats.

by congenital hydrocephalus which required sacrifice of one male in the third month. The animals with ataxia and abnormal aggression all had brain tumours, and the animals with paraplegia had severe muscle atrophy. No cause was found for excessive salivation in the histological examination of the salivary glands. Early sacrifice was considered important for animals with this condition in case they had viral sialoadenitis, but serum samples from affected animals were negative for rat corona virus. Two males and one female were sacrificed because of severe malocclusion which prevented normal feeding. General lethargy was related to renal disease, and neoplasia and laboured respiration to lung tumours. Pale extremities were confirmed, by haematology investigations, to be due to anaemia caused by haemorrhage or leukaemia. Only a few animals with masses died or had unscheduled necropsies; in these animals the mass had ruptured or ulcerated, or was so large as to restrict normal movement.

In Table 1.13 it is clear that skin lesions were the most common clinical observation, and they are discussed in greater detail in the chapter on the integumentary system. The majority of the skin lesions were thought to be caused by fighting or self-mutilation, as were tail abnormalities, which included abscesses and missing tail tips. Masses included neoplasms and non-neoplastic lesions such as skin cysts; these masses remained small and did not appear to affect the general health of the animals. Many of the observations are non-specific signs of general ill health such as piloerection, porphyrin staining around the eyes (chromodacryorrhea) and mild urinogenital staining, while distended abdomens were related to a range of conditions including faecal impaction and intra-abdominal masses.

Table 1.13 Incidence of less important clinical observations

Clinical observation	% Incidence[a]	
	Males	Females
Skin lesions		
Scabs, scratches, bites	25	12
Swollen feet	11	2
Swollen ears	1	0
Masses	16	28
Piloerection	3	4
Alopoecia	9	26
Missing whiskers	0	4
Urinogenital staining	8	3
Porphyrin staining	8	8
Tail abnormalities	20	3
Distended abdomen	2	18

[a] Incidence of clinical observations in animals surviving to termination of a 2 year study in 300/sex control AP rats.

1.12 Mortality

Mortality is, of course, dependant on many factors such as the health status of the animals and the diet they are fed, thus comparisons between laboratories are not of great use. On the other hand, comparisons within a colony are important, to detect any significant changes which may be occurring in the strain. The most important effect on mortality for the AP rat was the change from conventional to SPF conditions. Comparisons of the strain raised under conventional conditions with those from the newly built SPF unit were published by Paget and Lemon (1965). Mortality was significantly different in the SPF rats. At 12 months 5.7 per cent of the conventionally reared animals had died compared with 1 per cent of the SPF animals, and by 24 months the difference was greater with 42 per cent of the conventional animals dead compared with 21 per cent of the SPF group, i.e. 79 per cent of the SPF animals were alive at 2 years. The difference was due to a higher mortality from infectious diseases, particularly bronchopneumonia, in the conventionally reared animals. We had assumed that mortality in the SPF rat would remain relatively unchanged provided the integrity of the SPF unit was maintained, but mortality in the AP rat has increased over the last three decades. In the animals used for 2 year toxicology studies between 1988 and 1992, survival was 55 to 60 per cent in males and 50 to 66 per cent in females. This is nearly a 20 per cent reduction in survival over the 30 year period and can be attributed to a rise in mortality from tumours and renal disease. This increase in mortality has been shown to be related to *ad libitum* feeding and the consequent increase in body weight (Roe and Tucker, 1973; Tucker, 1979); it is hoped that the introduction of the maintenance diet R&M No 1 will modify this effect.

1.13   Histopathology

The histopathology of diseases in the AP rat was first investigated in a life-span study of 482 breeding and virgin male and female AP rats, which were kept for the whole of their life-span in the SPF breeding unit. Some of this work was reported by Paget and Lemon (1965). Since 1965, at approximately 5 year intervals, the incidence of spontaneous disease has been studied in groups of untreated rats (minimum 100/sex) which were maintained for 2 years in the experimental animal facility. These groups form part of the database which also includes the control animals from toxicology studies. The tissues examined have been similar throughout the period, although additional samples of some organs such as the liver, salivary glands and intestines have been added at various times. The standard tissue list for toxicology studies includes the following: adrenal glands, bone and marrow (sternum), brain, cervix, Harderian glands, heart, femur, intestines (duodenum, jejunum, ileum, caecum, colon), kidneys, liver, lungs, lymph nodes (mandibular, mesenteric), mammary gland, nerve (sciatic), oesophagus, ovaries, pancreas, pituitary, prostate, salivary glands (parotid, submandibular, sub-lingual), seminal vesicle, skin, spinal cord, spleen, stomach, testes and epidiymides, thymus, thyroid, trachea, urinary bladder, uterus, vagina. Additional samples are taken from any tissue which is macroscopically abnormal. Femoral bone marrow smears are taken at necropsy and stained by Giemsa's technique. In the first 15 years all tissues were fixed in Zenkers fixative, thereafter in 10 per cent buffered formalin, except for eyes which were fixed in Davidson's fluid. Tissues were processed by standard techniques and embedded in paraffin wax. Sections were cut at a nominal 5 µm and stained with haematoxylin and eosin (H&E).

The incidence of spontaneous diseases in the AP rat described in this book is confined to animals between 2 and 110 weeks of age. Many, if not all, diseases in the rat are progressive, so that incidence levels only reflect the picture at a specific time point. Where possible these will be compared with incidence levels in the life-span study.

1.14   References

ALT, J., HACKBARTH, H., DEERBERG, F. and STOLTE, H. (1980) Proteinurea in rats in relation to age dependant renal changes, *Laboratory Animals*, **14**, 95–101.
ARCHER, R. K. and RILEY, J. (1981) Standardised method for bleeding rats, *Laboratory Animals*, **16**, 25–8.

BAILLY, Y. and DUPRAT, P. (1990) Normal blood cell values in the rat, in JONES, T. C., WARD, J. M., MOHR, U. and HUNT, R. D. (Eds), *Haemopoietic System*, Berlin: Springer-Verlag.

CAMERON, T. P., HICKMAN, R. L., KORNREICH, M. R. and TARONE, R. E. (1985) History, survival and growth patterns of B6C3F1 mice and F344 rats in the National Cancer Institute carcinogenesis testing program, *Fundamental and Applied Toxicology*, **5**, 526–38.

CHARLES RIVER INC. (1989) *Historical Data Base.*

COLEMAN, G. L., BARTOLD, S. W., OSBALDISTON, G. W., FOSTER, S. J. and JONAS, A. M. (1977) Pathological changes during aging in barrier-reared Fischer F344 male rats, *Journal of Gerontology*, **32**, 258–78.

DAVEY, D. G. (1959) Establishing and maintaining a colony of Specific Pathogen free mice, rats and guinea pigs. Symposium on the quality of laboratory animals, *Laboratory Animal Centre Collected Papers*, **8**, 17–34.

FESTING, M. (1974) Genetic monitoring of laboratory mouse colonies in the Medical Research Council Accreditation Scheme for the suppliers of Laboratory Animals, *Laboratory Animals*, **8**(3), 291–99.

HORBACH, G., YAP, S. H. and VAN BEEZOOIJEN, C. F. (1983) Age related changes in albumin elimination in female WAG/Rij rats, *Biochemical Journal*, **216**, 309–15.

KLINGER, M. M., MACCARTER, G. D. and BOOZER, C. N. (1996) Body weight and composition in the Sprague-Dawley rat: comparison of three outbred sources, *Laboratory Animal Science*, **16**, 67–70.

LANG, P. L. and WHITE, W. J. (1992) Growth, development, and survival of the Crl:CD®(SD)BR stock and CDF®(F344/CriBR) strain, in MOHR, U., DUNGWORTH, D. L. and CAPEN, C. C. (Eds), *Pathobiology of the Aging Rat*, Vol. 2, pp. 587–608, Washington: ILSI Press.

LEONARD, R. and RUBEN, Z. (1986) Hematology reference values for peripheral blood of laboratory rats, *Laboratory Animal Science*, **36**, 277–81.

LEWI, P. J. and MARSBOOM, R. P. (1981) *Toxicology reference data – Wistar rat. Body and organ weights, biochemical determinations, haematology and urinalyses*, Amsterdam: Elsevier.

LOEB, W. F. and QUIMBY, F. W. (1989) *The Clinical Chemistry of Laboratory Animals*, New York: Pergamon.

MASDEN, N. B. and TUBA, J. (1952) On the source of the alkaline phosphatase in rat serum, *Journal of Biological Chemistry*, **195**, 741–50.

PAGET, G. E. and LEMON, P. G. (1965) The interpretation of pathology data, in RIBELIN, W. E. and MCCOY, J. R. (Eds), *The Pathology of Laboratory Animals*, pp. 382–405, Springfield, Illinois: Charles C. Thomas.

PARKENING, T. A., COLLINS, T. J. and SMITH, E. R. (1983) Measurement of plasma LH concentration in aged male rodents by a radio-immunoassay and a radioreceptor assay, *Journal of Reproduction and Fertility*, **69**, 717–22.

PAYNE, B. J., LEWIS, H. B., MURCHISON, T. E. and HART, E. A. (1976) Hematology of laboratory animals, in MELBY, E. D. and ALTMAN, N. H. (Eds), *Handbook of Laboratory Animal Science*, Vol. III, pp. 383–61, Cleveland, Ohio: CRC Press.

ROE, F. J. C. and TUCKER, M. J. (1973) Recent developments in the design of carcinogenicity tests on laboratory animals, *Proceedings of the European Society for the Study of Drug Toxicity*, **XV**, 171–7.

SUBER, R. L. and KODELL, R. L. (1985) The effect of phlebotomy techniques on hematological and clinical evaluation in Sprague-Dawley rats, *Veterinary and Clinical Pathology*, **14**, 23–30.

TUCKER, M. J. (1979) The effect of long term food restriction on tumours in rodents, *International Journal of Cancer*, **23**, 803–7.

UPTON, P. K. and MORGAN, D. J. (1975) The effect of sampling on some blood parameters in the rat, *Laboratory Animals*, **9**, 85–91.

WEIL, C. S. (1982) Statistical analyses and normality of selected hematologic and clinical chemistry measurements used in toxicologic studies, *Archives of Toxicology*, **5**, 237–53.

YAMANUCHI, C., FUJITA, S., OBARA, T. and VEDA T. (1981) Effects of room temperature on reproduction, body and organ weights: food and water intake and hematology in rats, *Laboratory Animal Science*, **31**, 251–8.

YARRINGTON, J. T. and JOHNSTON, J. O'N. (1994) Ageing in the adrenal cortex, in MOHR, U., DUNGWORTH, D. L. and CAPEN, C. C. (Eds), *Pathobiology of the Aging Rat*, Vol. 2, pp. 227–44, Washington: ILSI Press.

YOUNG, G. P., FRIEDMAN, S., YEDLIN, S. T. and ALPERS, D. H. (1981) Effect of fat feeding on intestinal alkaline phosphatase activity in tissue and serum, *American Journal of Physiology*, **241**, G461–G468.

ZAGER, R. A. and ALPERS, C. E. (1989) Effects of aging on expression of ischaemic acute renal failure in rats, *Laboratory Investigation*, **61**, 290–4.

2

The Integumentary System and Mammary Glands

2.1 Skin

The skin is a complex tissue with a variety of components (Figure 2), including the epidermis, epidermal appendages such as sebaceous glands and hair follicles, and the dermis. In the rat only the skin of feet, mouth and snout are free of fur, and in albino rats although melanocytes are present in the skin the pigment melanin is totally absent. The development and structure of the skin in the albino rat has been described in detail by English and Munger (1994). Disorders in other tissues can be associated with changes in the skin, but skin lesions are generally less common in the rat than in humans.

2.1.1 Non-neoplastic Changes

Inflammation

In the AP rat, minor abrasions of the skin have occurred as the result of trauma from bites, scratches or contact with metal parts of cages. They may result in erosions or ulceration and scab formation. Inflammatory conditions may be accompanied by epidermal changes including acanthosis, hyperkeratosis and parakeratosis. The incidence in AP rats is similar in males and females and can reach 20 per cent at 2 years. In long term studies swollen feet, with oedema and dermal cellulitis, is more common in males, reaching a maximum incidence of 15 per cent in males and 4 per cent in females. These changes may be related to changes in the sweat (eccrine) glands which are only present in the footpads of rats. Atrophy of these glands may be due to long term contact with the mesh floors of cages, and the higher incidence in males may be attributed to the larger body weight and consequent increased pressure on the feet of males.

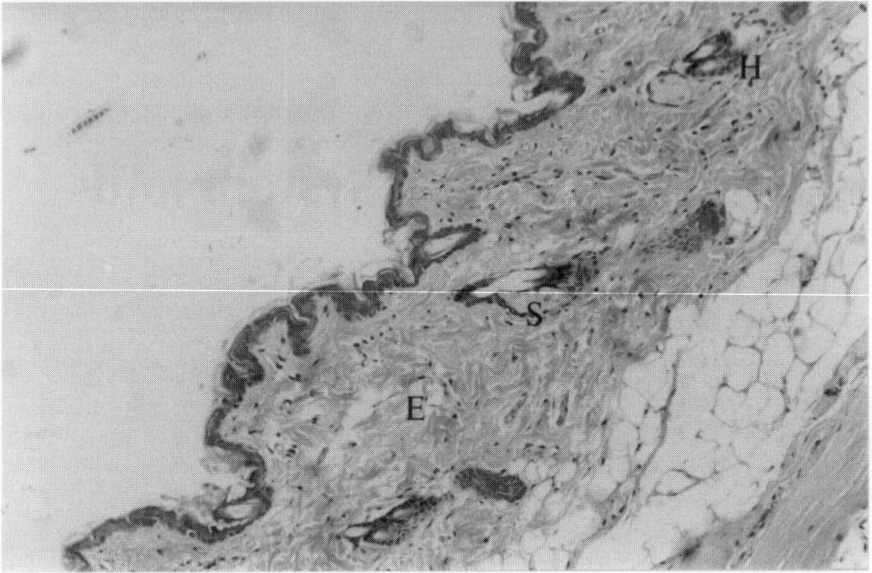

Figure 2   Normal skin of a male AP rat showing epidermis (E), hair follicles (H) and sebaceous glands (S). ×8, haematoxylin and eosin (H&E)

Alopecia

Alopecia can occur in association with these skin lesions; it can also be related, rarely, to viral or bacterial infections of the skin. In the AP rat it occurs most frequently without any obvious histological change in the skin, other than atrophy of hair follicles. This type of idiopathic hair loss is more common in female AP rats and can reach an incidence of 35 per cent at 2 years. Rodent hairs are normally replaced in waves across the body, unlike human hair which is replaced in a random fashion (Butcher, 1950; Montagna and Parakkal, 1974). The factors which control the wave replacement are not known. Missing whiskers are not common, and this hair loss probably results from grooming since histological sections of skin from animals with missing whiskers show no changes in the vibrissae other than the missing hair shaft.

Tail

In animals more than 18 months old a missing tail tip has been observed in up to 8 per cent of males and this is most likely to be related to tail chewing, which has also been observed only in males. The skin of the tail is smooth at birth but scale formation begins rapidly, so that after 3 weeks there are 45 rings of scales and at 1 year 190 rings (Erickson, 1931). The macroscopic appearance of 'ringtail' is of a tail with transverse bands of constriction at regular intervals along the tail, and the microscopic appearance is of rings of eroded or

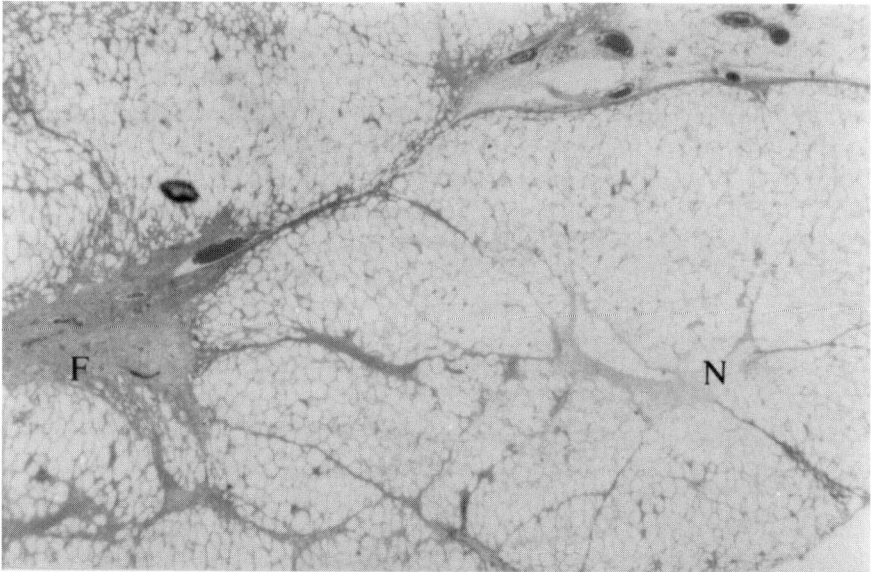

Figure 3   Omental fat from an AP rat showing necrosis (N) and fibrosis (F). ×8, H&E

ulcerated, acanthotic and hyperkeratotic epidermis with a marked dermatitis. This has been seen occasionally in both breeding and experimental AP rats and is thought to be related to changes in environmental humidity.

Cysts

A common change in the skin is the presence of cysts, which are chiefly epidermal cysts, lined by a simple squamous epithelium and filled with keratin; occasionally a more florid epithelium with some sebaceous elements may be seen. Dermal cysts are rare and diagnosis requires the presence of some dermal feature such as hair or hair follicles. Rupture of epidermal cysts, with escape of keratin into the adjacent tissues, may cause a localised foreign body or granulomatous reaction. It is difficult to be accurate about the incidence of such cysts as they are usually only detected when they are quite large; the highest incidence recorded in a 2 year study was 5 per cent.

Fat necrosis

Necrosis of adipose tissue is most commonly seen in infarcted omental fat which undergoes patchy necrosis and fibrosis (Figure 3). The fibrotic nodule may become detached and form a free floating cream/yellow coloured body in the abdominal cavity. This change has been seen in several strains (Snell, 1965).

2.1.2 Neoplastic Changes

Primary skin tumours are uncommon in young (less than 12 months old) AP rats, and the incidence varies markedly among different strains. They have been reviewed by Kovatch (1990) in the F344 rat and in various other strains (Zackheim, 1973; Burek, 1978a; Zwicker *et al.*, 1994). In the AP rat the overall incidence in control groups in 2 year studies varies from 0 to 7.5 per cent. The incidence in males (maximum 15 per cent) is higher than in females (3 per cent). The most common sites for tumours in descending order of frequency are in the skin of the head, abdomen and feet.

Epithelial tumours

The earliest tumour observed was a squamous papilloma in a male aged 2 months but, in general, they are found in old animals, the majority during scheduled necropsy at the end of a 2 year study. The incidence of epithelial tumours is shown in Table 2.1. Squamous cell tumours are the most common type of skin tumour, with papillomas more common than carcinomas. In the AP rat both benign and malignant tumours are well differentiated, with the presence of keratin providing a diagnostic feature. Keratoacanthomas are very rare in AP

Table 2.1 Incidence of epithelial tumours of skin in the AP rat

	% Incidence	
Tumour[a]	Males	Females
Squamous papilloma	13	1
Squamous carcinoma	4	1
Keratoacanthoma	1[b]	0
Basal cell carcinoma	3	0
Tricholepithelioma	2[b]	0
Baso-squamous tumours	2	2
Sebaceous cell adenoma	1	0
Sebaceous cell carcinoma	1[b]	0
Squamous sebaceous adenoma	3	3
Squamous sebaceous carcinoma	1	0
Amelanotic malignant melanoma	1[c]	0

[a] Incidence of skin tumours derived from a database of 8880 control animals (including 2800 males and 2500 females in 2 year studies) used in studies between 1960 and 1992. Incidence levels are the highest recorded in any of 24 oncogenicity studies. A zero incidence has been recorded for every tumour type in at least 5 studies.
[b] Low overall incidence, only two in the database.
[c] Low overall incidence, only one in the database.

rats, unlike SD rats where incidences up to 4 per cent have been recorded (Zwicker *et al.*, 1994). All other types are rare in the AP rat and include basal cell carcinomas, sebaceous cell adenomas and carcinomas, and tumours which show a mixture of cell types such as squamous sebaceous cell adenomas, basosquamous adenomas and carcinomas. These tumours were all well differentiated, with easily identified elements of the cells of origin. Metastases from squamous carcinomas have been found in the lung in two animals. Two trichoepitheliomas and a single amelanotic malignant melanoma of the skin have been observed (the melanoma was diagnosed by ultrastructural examination).

Mesenchymal tumours

Mesenchymal tumours of the skin are among the most common in the AP rat, with the total incidence ranging between 0 and 15 per cent in males and 0 and 5 per cent in females. The highest and lowest incidences of individual types are shown in Table 2.2. Subcutaneous fibromas are the most common

Table 2.2 Incidence of mesenchymal tumours of skin in the AP rat

Tumour	% Incidence[a]	
	Males	Females
Fibroma	15	8
Fibrosarcoma	4	2
Neurofibroma[b]	1	0
Fibrous histiocytoma		
benign	1	1
malignant	4	1
Lipoma	12	6
Liposarcoma	1	1
Hibernoma[b]	1	1
Angioma	2	1
Angiosarcoma	1	1
Leiomyosarcoma[b]	1	0

[a] Tumour incidence derived from a database of 8880 control animals (including 2800 males and 2500 females in 2 year studies) used in studies between 1960 and 1992. Incidence levels are the highest recorded in any of 24 oncogenicity studies. The lowest incidence for all tumours was zero except fibromas, where the lowest recorded incidence was 4%.

[b] Low overall incidence – 2 neurofibromas, 2 hibernomas, 1 leiomyosarcoma.

27

mesenchymal tumour in the AP rat. Diagnostic features are dense bands of collagen with few fibroblastic cells and a low mitotic activity. Since they are benign they are usually found in animals killed at the termination of long term studies. Food restriction reduces the incidence of this type of tumour (Tucker, 1979). Fibromas are also the most common mesenchymal tumour of the SD rat (Zwicker *et al.*, 1994). Fibrosarcomas are less frequent, but also most common in animals over 20 months, although the earliest seen in the AP rat was in an animal 1 month old. They are more cellular than the benign tumour and show less collagen. Considerable pleomorphism is usually a feature and there may be marked mitotic activity. Metastases to the lung have been observed with three fibrosarcomas. Fibrohistiocytic tumours have a variable histological appearance, distinct from fibrosarcomas, and have been described in detail by Greaves and Faccini (1981). They may be highly pleomorphic with mixtures of histiocytic cells, spindle cells, multinucleate cells and giant cells; vascular components may be prominent. Malignant fibrohistiocytic tumours are more common than the benign variety. They show widespread local invasion and occasional metastases to lymph nodes and lung. They were not reported by Zwicker *et al.* (1994) in their comparison of F344, SD and Wistar rats. Neurofibromas are rare in the AP rat and also in other colonies of Wistar rats, but an incidence of 8 per cent in males was recorded in F344 rats by Zwicker *et al.* (1994). Lipomas and liposarcomas are uncommon in all strains (Carter, 1973; Anver, 1989a, 1989b; Zwicker *et al.*, 1994). The benign tumours are easily diagnosed by the presence of mature fat cells separated by fibrous septa (Figure 4), while the liposarcomas, although always containing some fat cells, have a mixed cell population with spindle cells and vascular elements and numerous mitotic figures. Hibernomas (tumours of brown fat) are another rare type of tumour in the rat, which has been reviewed by Coleman (1989). The two hibernomas in the AP rat were both in the mediastinum, and one showed extensive infiltration into the lung (Figures 5 and 6). Vascular neoplasms are not uncommon in the AP Wistar rat, but are more frequent in the lymph node than in the subcutis. A small incidence was also seen in the three rat strains reported by Zwicker *et al.* (1994). The only dermal leiomyosarcoma was seen in a male aged 26 months. The tumour was well differentiated and showed only minimal local invasion.

2.2 Zymbal's Glands

2.2.1 *Non-neoplastic Changes*

Inflammation

The auditory sebaceous glands of the rat (Zymbal's glands) have not been routinely examined in the AP rat, but only when macroscopic abnormalities

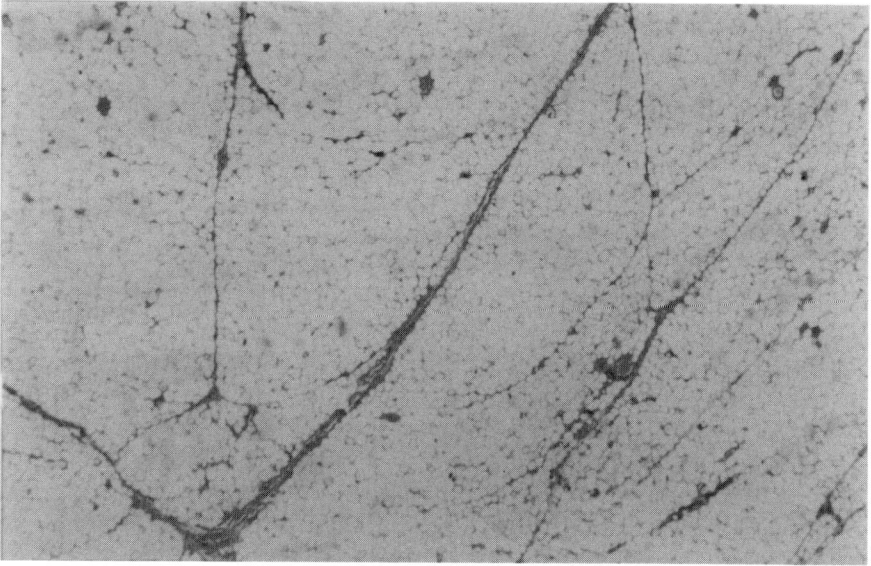

Figure 4 Well-differentiated lipoma from an AP rat, showing mature fat cells traversed by fibrous strands. ×8, H&E

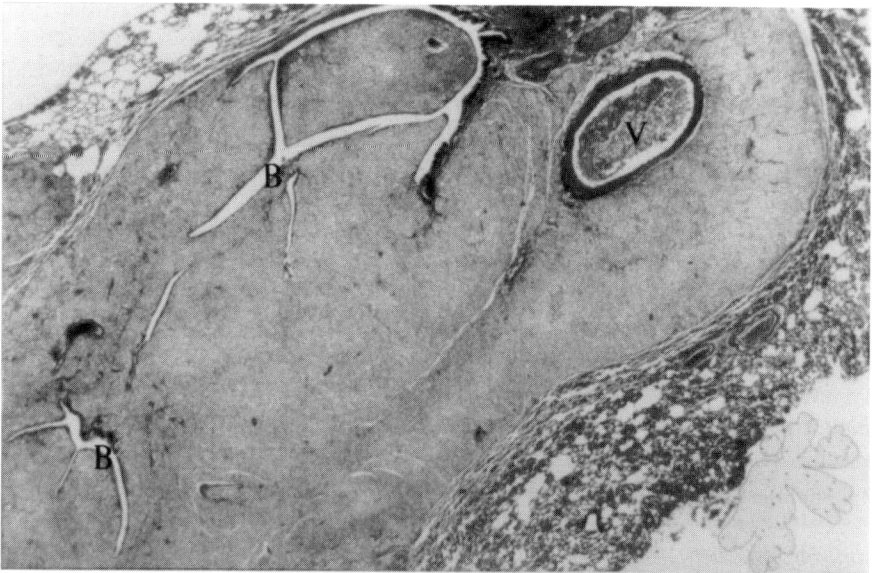

Figure 5 Hibernoma in the lung of an AP rat. The tumour is surrounding the bronchi (B) and blood vessels (V). ×32, H&E

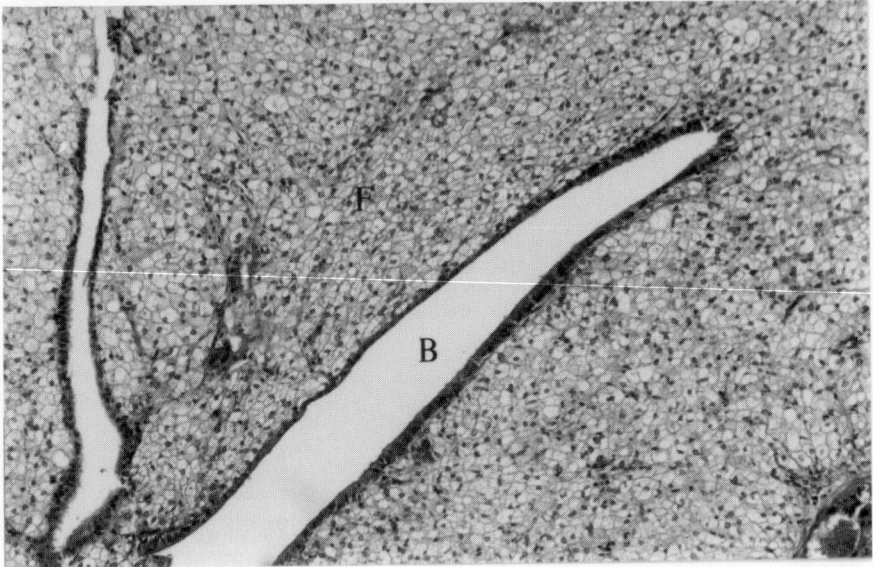

Figure 6   Hibernoma of mature brown fat cells (F) surrounding bronchi (B). ×32, H&E

were observed. The function of these bilateral glands, which are situated near the external ear canals, is uncertain. It has been suggested that, like the preputial gland, they may produce pheromones or, as they exhibit cytochrome P-450 enzyme activity, they may have some metabolic function (Pohl and Fouts, 1983). They have only been examined in the AP rat when macroscopic abnormalities have been detected. Few changes have been seen, and these include small mononuclear cell infiltrates, acute and chronic adenitis with some ductal ectasia.

2.2.2   *Neoplastic Changes*

Tumours of the glands appear very rarely in the AP rat, with only three adenocarcinomas in the database. This may be an inaccurate figure since tumours may not be sufficiently large to produce a macroscopic abnormality, in which case the gland would not have been examined histologically. Since most toxicologists do not take routine sections through the ears, this probably accounts for the low spontaneous incidence recorded for most strains. Spontaneous and induced tumours of the gland have been reviewed by Pliss (1973). Zymbal's gland tumours have been induced in the AP rat with 2-acetylaminofluorene (unpublished data).

2.3   Preputial Glands

2.3.1   *Non-neoplastic Changes*

Inflammation

The preputial gland has also not been examined routinely in the AP rat, but only when macroscopic abnormalities were present. Inflammatory lesions, including acute and chronic adenitis and abscesses, have been seen with cystic dilatation of ducts.

2.3.2   *Neoplastic Changes*

A single spontaneous squamous cell carcinoma has been found in the preputial gland, but this also cannot be considered an accurate indication of incidence since the gland has been examined so infrequently. A high incidence of 8.7 per cent has been recorded in F344 rats (Coleman *et al.*, 1977) although lower incidences have been seen in other strains (Hiraga and Fujii, 1977).

2.4   Mammary Glands

2.4.1   *Non-neoplastic Changes*

In terms of morbidity and mortality the mammary gland is a very important tissue in all strains of rat. The anatomy, histology and development of the gland has been described in detail by Russo and Russo (1994). Briefly, there are six pairs of mammary glands in the rat on the ventral/lateral surface, with the first pair in the cervical area. Mammary tissue in these glands can extend inwards to the salivary glands and, in older animals when the mammary tissue is hyperplastic, it is difficult to separate salivary glands from cervical mammary tissue. There are a further two thoracic pairs, an abdominal pair, and two pairs in the inguinal area. In the AP rat the routine sample of mammary tissue in toxicology studies is taken from one of the left inguinal glands.

Inflammation and hyperplasia

Minimal chronic and acute inflammation has been observed in virgin rats, and occasional abscesses or acute necrotising mastitis in breeding females. The most common non-neoplastic conditions are ectasia and hyperplasia. In the AP rat, mammary hyperplasia is a progressive condition in both sexes, although in males changes are less frequent and less severe. In females less than 6 months

31

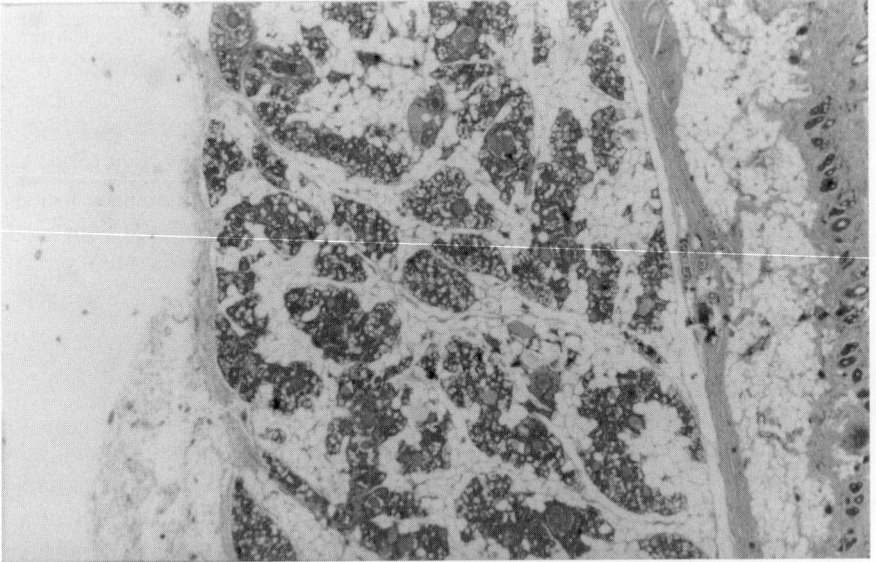

Figure 7 Moderate mammary gland hyperplasia in an 18 month old female AP rat showing acinar hyperplasia. ×8, H&E

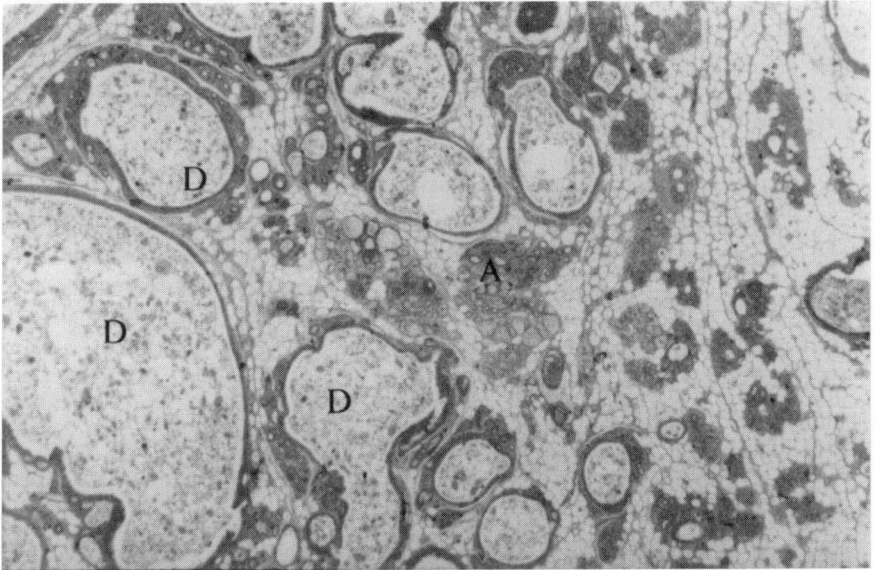

Figure 8 Mammary gland hyperplasia showing dilated ducts (D) and hyperplastic acini (A). ×32, H&E

of age there is no significant hyperplasia, although the glands enlarge to varying degrees as part of normal growth. At 12 months lobular hyperplasia, i.e. hyperplasia of the acini (alveoli) is apparent, accompanied by ectasia of ducts with a little hyperplasia of duct epithelium; in some the dilatation of ducts is so great that the term galactocoele is used. These ducts are filled with eosinophilic material. By 18 months most females will show moderate to severe hyperplasia (Figures 7 and 8). The incidence of hyperplasia varies with strain, ranging from a relatively low incidence in F344 rats (Goodman *et al.*, 1979) and Osborne-Mendel rats (Goodman *et al.*, 1980) to 100 per cent incidence in the WAG/Rij (Burek, 1978b) and in the female AP rat at the end of 2 year studies.

2.4.2 Neoplastic Changes

The different types of mammary tumour observed in AP rats are given in Table 2.3. Fibroadenomas are the most common type observed. These tumours are composed of well-differentiated fibrous and epithelial tissues, although they are not present in equal proportions, with some tumours having few epithelial areas and others with little collagen. Multiple tumours are common in the AP rat, but the most common site is the inguinal glands followed by the thoracic glands. The tumours can be very large, and if multiple tumours are present it may be difficult to separate the individual tumours. Since they are benign tumours they do not normally kill the animal unless the surface of the tumour is damaged by physical trauma and becomes infected, or if the size of the tumour causes difficulty in movement, or if general body condition deteriorates. Only a few

Table 2.3 Incidence of mammary tumours in AP rats

Mammary tumour	Male Highest	Lowest	Female Highest	Lowest
	% Incidence[a]			
Fibroadenoma	1.7	0	40.0	10.4
Adenoma	0.3	0	2.0	0
Fibroma	0	0	1.0	0
Adenocarcinoma	0.4	0	19.0	6.0
Carcinosarcoma	0	0	1.4	0

[a]Tumour incidence derived from the database of 8880 control animals (including 2800 males and 2500 females in 2 year studies) used in toxicology studies between 1960 and 1994. Incidence levels are the highest and lowest percentage observed in the 24 oncogenicity studies.

adenomas have been observed and these were all relatively small masses composed almost entirely of glandular tissue with a minimal connective tissue stroma. The glandular epithelium was, typically, composed of uniform alveoli lined by a single cell layer. Fibromas have the histological appearance described for the subcutaneous tumour but they arise within the mammary gland. Since fibroadenomas may have areas which are fibrotic and devoid of epithelial elements, there is no histological feature to distinguish mammary fibromas from fibroadenomas unless adequate samples are taken for examination. The macroscopic appearance of the tumours is different, with fibroadenomas most often having a distinct lobulated pattern to the cut surface, and they may exude a milky fluid. Fibromas have a more regular cut surface and are firm and tough to cut and do not appear lobulated.

Adenocarcinomas exhibit a wide range of histological patterns including alveolar and papillary epithelium and dense masses of poorly differentiated cells. Mitotic figures are usually frequent and connective tissue sparse. In some tumours metaplasia to squamous or sebaceous cells has been observed, and in some the epithelial cells are so pleomorphic and undifferentiated that they have been described as anaplastic carcinomas. Greaves and Faccini (1984) considered the undifferentiated cells represent a myoepithelial component of the tumours. Metastases to the lung have been seen in only a few animals. A notable feature of the incidence of mammary tumours in female AP rats is that, although the overall incidence has doubled from levels of 20 per cent to 40 per cent in the 35 years since the colony was established at Alderley Park, the change in the ratio of benign to malignant tumours has also altered. Between 1960 and 1975 the ratio of benign to malignant mammary tumours was 10:1 but, thereafter, the incidence of malignant tumours increased and the ratio has reached 2:1. The reason for this change is not known but it is not related to changes in diagnostic criteria, as all mammary tumours have been reviewed. The development of mammary tumours is dependant on the presence of the two hormones prolactin and oestrogen, and it has been suggested (Clifton, 1979) that it is the ratio of these hormones that is critical, not the absolute levels. It is well established that increased prolactin levels, produced by a variety of methods, e.g. pituitary tumours, oestrogen administration or dopamine inhibitors, will increase mammary tumours (Welsch and Nagasawa, 1977) and prolactin reduced by any method inhibits the development. Diet also influences the development of tumours, and in the AP rat food restriction of 20 per cent reduced the overall incidence of mammary tumours five fold in a 2 year study (Tucker, 1979). In that study the ratio of benign to malignant tumours in the *ad libitum* group was approximately 3:1 (17 benign and 6 malignant), and a similar ratio was found in the dietary restricted group (three benign and one malignant). Pituitary tumours in the AP rat have also increased over the years (see Chapter 12 on the endocrine system) and these have been shown to be prolactin-secreting tumours. This is the probable cause of the increased overall incidence of mammary tumours but not the increased proportion of malignant tumours.

2.5 References

ANVER, M. R. (1989a) Lipoma, subcutis, rat, in JONES, T. C., MOHR, U. and HUNT, R. D. (Eds), *Pathology of Laboratory Animals*, pp. 100–3, New York: Springer-Verlag.

ANVER, M. R. (1989b) Liposarcoma, subcutis, rat, in JONES, T. C., MOHR, U. and HUNT, R. D. (Eds), *Pathology of Laboratory Animals*, pp. 103–6, New York: Springer-Verlag.

BUREK, J. D. (1978a) Skin and subcutaneous tissues, in *Pathology of Aging Rats*, pp. 161–3, Boca Raton, Florida: CRC Press.

BUREK, J. D. (1978b) Non-neoplastic and neoplastic lesions of the mammary gland, in *Pathology of Aging Rats*, pp. 163–8, Boca Raton, Florida: CRC Press.

BUTCHER, E. O. (1950) Development of the pilary system and the replacement of hair in mammals, *Annals of the New York Academy of Sciences*, **53**, 508–16.

CARTER, R. L. (1973) Tumours of the soft tissues, in TURUSOV, V. S. (Ed.), *Pathology of Tumours in Laboratory Animals*, Vol. 1, *Tumours of the Rat*, pp. 151–68, Lyon: IARC.

CLIFTON, K. H. (1979) Animal models of breast cancer, in ROSE, D. P. (Ed), *Endocrinology of Cancer*, Vol. 1, pp. 1–20, Boca Raton, Florida: CRC Press.

COLEMAN, G. L. (1989) Hibernoma, rat, in JONES, T. C., MOHR, U. and HUNG, R. D. (Eds), *Integument and Mammary Gland*, pp. 126–9, New York: Springer-Verlag.

COLEMAN, G. L., BARTOLD, S. W., OSBALDISTON, G. W., FOSTER, S. J. and JONAS, A. M. (1977) Pathological changes during aging in barrier-reared F344 rats, *Journal of Gerontology*, **32**, 258–78.

ENGLISH, K. B. and MUNGER, B. L. (1994) Normal development of the skin and subcutis of the albino rat, in MOHR, U., DUNGWORTH, D. L. and CAPEN, C. C. (Eds), *Pathobiology of the Aging Rat*, Vol. 2, pp. 363–89, Washington: ILSI Press.

ERICKSON, T. C. (1931) The post natal development of the caudal integument in the rat, *American Journal of Anatomy*, **47**, 173–93.

GOODMAN, D. G., WARD, J. M., SQUIRE, R. A., CHU, K. C. and LINHART, M. S. (1979) Neoplastic and non-neoplastic lesions in aging F344 rats, *Toxicology and Applied Pharmacology*, **48**, 237–48.

GOODMAN, D. G., WARD, J. M., SQUIRE, R. A., CHU, K. C. and LINHART, M. S. (1980) Neoplastic and non-neoplastic lesions in aging Osborne-Mendel rats, *Toxicology and Applied Pharmacology*, **55**, 433–47.

GREAVES, P. and FACCINI, J. M. (1981) Spontaneous fibrous histiocytic neoplasms in rats, *British Journal of Cancer*, **43**, 402–11.

GREAVES, P. and FACCINI, J. M. (1984) Integumentary system. In *Rat Histopathology*, pp. 8–34, Amsterdam: Elsevier.

HIRAGA, K. and FUJII, T. (1977) Tumours of the preputial gland in rats, *Gann*, **68**, 369–70.

KOVATCH, R. M. (1990) Neoplasms of the integument, in STINTON, S. F., SCHULLER, H. M. and REZNIK, G. (Eds), *Atlas of Tumor Pathology of the Fischer Rat*, pp. 20–32, Boca Raton, Florida: CRC Press.

MONTAGNA, W. and PARAKKAL, P. F. (1974) *The Structure and Function of Skin*, New York: Academic Press.

PLISS, G. B. (1973) Tumours of the auditory sebaceous glands, in TURUSOV, V. S. (Ed.), *Pathology of Tumours in Laboratory Animals*, Vol. 1, *Tumours of the Rat*, pp. 23–30, Lyon: IARC.

POHL, R. J. and FOUTS, J. R. (1983) Cytochrome P-450-dependant xenobiotic metabolising activity in Zymbal's gland, a specialised sebaceous gland of rodents, *Cancer Research*, **43**, 3660–2.

RUSSO, I. H. and RUSSO, J. (1994) Aging of the mammary gland, in MOHR, U., DUNGWORTH, D. L. and CAPEN, C. C. (Eds), *Pathobiology of the Aging Rat*, Vol. 2, pp. 447–58, Washington: ILSI Press.

SNELL, K. C. (1965) Spontaneous lesions of the rat, in RIBELIN, W. E. and MCCOY, J. R. (Eds), *The Pathology of Laboratory Animals*, pp. 241–302, Springfield, Illinois: Charles C. Thomas.

TUCKER, M. J. (1979) The effect of long term food restriction on tumours in rodents, *International Journal of Cancer*, **23**, 803–7.

WELSCH, C. W. and NAGASAWA, H. (1977) Prolactin and murine mammary tumorigenesis: a review, *Cancer Research*, **37**, 951–63.

ZACKHEIM, H. S. (1973) Tumours of the skin, in TURUSOV, V. S. (Ed.), *Pathology of Tumours in Laboratory Animals*, pp. 1–22, Lyon: IARC.

ZWICKER, G. M., EYSTER, R. C., SELLS, D. M. and GASS, J. H. (1994) Comparative incidences of skin neoplasms in Sprague-Dawley, F344 and Wistar rats, in MOHR, U., DUNGWORTH, D. L. and CAPEN, C. C. (Eds), *Pathobiology of the Aging Rat*, Vol. 2, pp. 391–421, Washington: ILSI Press.

3

The Musculo-skeletal System

3.1 Muscle

In humans skeletal muscle accounts for approximately 40 per cent of body weight, but diseases of muscle are uncommon in the rat. The left quadriceps is the only muscle taken for histological examination in regulatory toxicology studies in the AP rat, although a range of muscles have been examined in other investigative studies. There are different types of muscle fibres and they respond differently to metabolic changes and workload (Lawrence *et al.*, 1986; Abe *et al.*, 1987). The slow twitch, Type 1 fibres have a low glycolytic activity and a high oxidative activity, and the fast twitch, Type 2 fibres the reverse: i.e. a high glycolytic activity and a low oxidative activity (Eddinger *et al.*, 1986; Ansved and Larsson, 1989). Endocrine control of muscle growth includes pituitary growth hormone, thyroid hormones and insulin, which all have growth-promoting effects, while glucocorticoids have catabolic effects.

3.1.1 *Non-neoplastic Changes*

Atrophy

The most important disease of the muscle in the AP rat is atrophy. The histological appearance is characterised by a reduction in the numbers of muscle fibres, with great variation in the size of the fibres, vacuolation, degeneration and fragmentation (Figures 9 and 10). It is usually thought that it is primarily Type 2 fibres which are affected, but Eddinger *et al.* (1985) have shown that this is not true in all rat strains and not true of all muscles. Clinically, when the disease is severe, the rats may be paraplegic, or in less

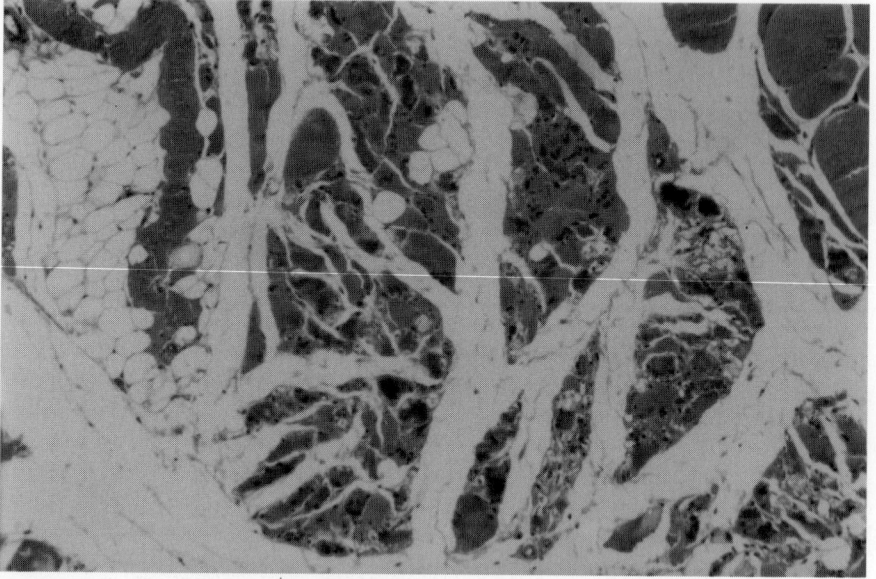

Figure 9   Muscle atrophy in a 26 month old male AP rat. ×32, H&E

severe cases the animals are reluctant to move and tend to 'drag' their hind legs. It is not a common disease and it is more frequently seen in males. In 2 year studies the incidence of paraplegia and severe muscle atrophy in males is approximately 5 per cent, but the incidence rises in older animals, and in the life-span study 40 per cent of males over 30 months of age showed extensive muscle atrophy. Although severe atrophy is rare, most rats at 2 years will show an occasional atrophic fibre. The aetiology of the disease is uncertain. AP rats at 2 years show extensive changes in the spinal nerves and nerve roots, and the incidence of these changes is much greater than that of muscle atrophy. Everitt *et al.* (1985) showed that muscle atrophy can be retarded by food restriction and hypophysectomy. In the AP rat it occurs in male animals with severe renal disease, some of which have minimal changes in nerves. This would suggest that there may be several factors involved in the disease and that changes in the nervous system may be only one factor.

Necrosis and inflammation

Most other muscle changes are rare in the AP rat and also in other strains. Necrosis of occasional muscle fibres occurs rarely (<1 per cent) and is usually unassociated with inflammation. Inflammation usually takes the form of small mononuclear inflammatory cell infiltrates between fibres and is as rare as necrosis. When arteritis is a widespread disease it may be seen in small vessels within the muscle; when present it has not been associated with muscle atrophy.

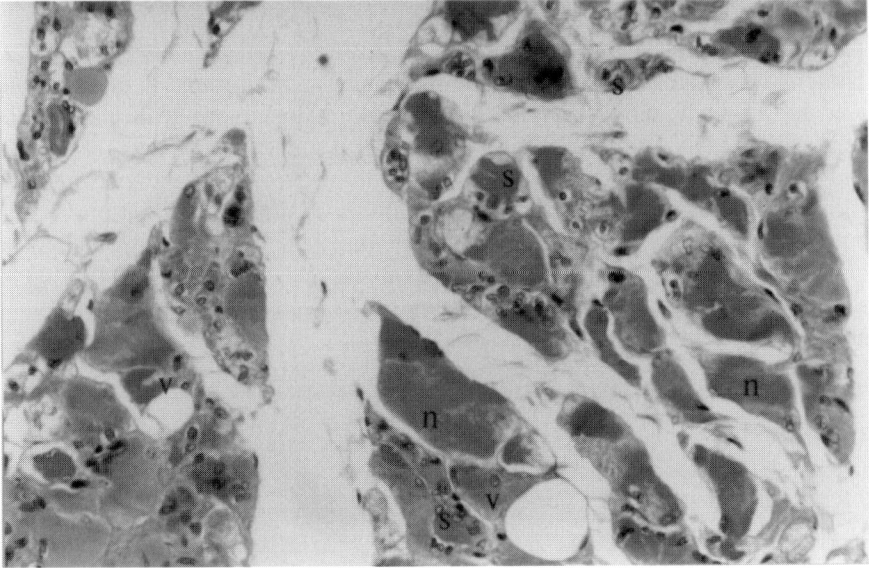

Figure 10 Muscle atrophy showing some normal fibres (n), vacuolated fibres (v) and shrunken fibres (s). ×128, H&E

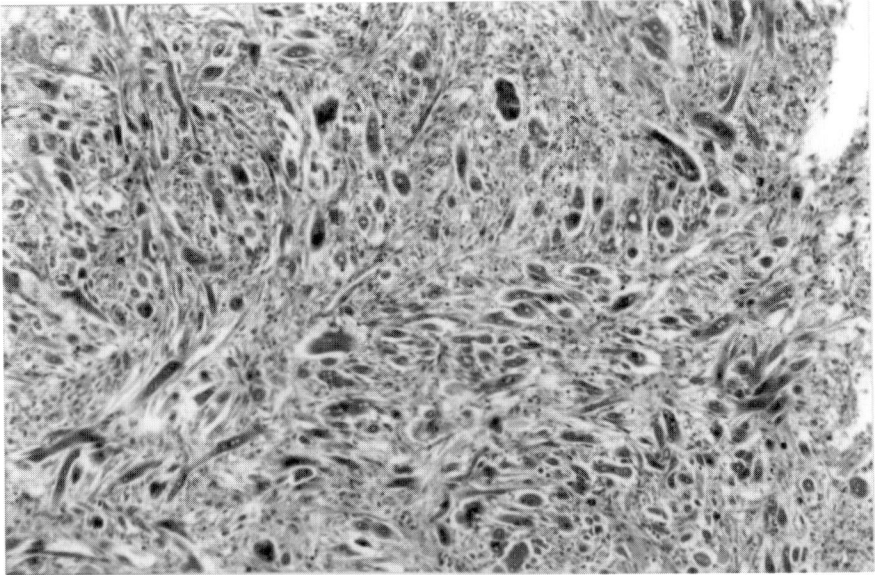

Figure 11 Rhabdomyosarcoma in a male AP rat showing the characteristic pleomorphic appearance with numerous giant cells. ×80, H&E

3.1.2   Neoplastic Changes

Tumours of skeletal muscle are rare in all strains of rat, including the AP rat. There are two rhabdomyosarcomas in the database, both found in males in the hind limbs. They show the typical pleomorphic appearance with many giant cells (Figure 11). Sections of the tumours were stained by various techniques and both showed cross striations, in some cells, when stained with phosphotunstic acid and haematoxylin. Secondary tumours of muscle are rare in AP rats but include angiosarcomas, osteosarcoma, leukaemias and malignant lymphomas. Reznik *et al.* (1980) found no muscle tumours in 60 048 F344 rats, and none was found in 1800 Wistars (Al Zubaidy and Malinowski, 1984) nor in 8960 SD rats (Krinke *et al.*, 1985).

3.2   Joints

3.2.1   Non-neoplastic Changes

Inflammation and degeneration

The left femoro-tibial joint is inspected in AP rats in short term (less than 2 years) toxicology studies, but histological examination of these joints only occurs when there is macroscopic abnormality. Few changes have been recorded in the joints. A minor degenerative change in the articular cartilage is not uncommon but it does not proceed to erosion and osteoarthritis. This is in agreement with the observations of Smale *et al.* (1995) who found that only minimal changes occurred in the articular cartilages of Wistar rats, compared with a 100 per cent incidence in F344 rats at all ages, and the degeneration increased with age. Inflammatory lesions of the joint have not been seen as a spontaneous condition although adjuvant arthritis is readily induced in the AP rat.

3.2.2   Neoplastic Changes

No primary neoplasms have been identified in a joint although a few large soft tissue tumours have invaded the joints.

3.3   Bone

3.3.1   Non-neoplastic Changes

Bone disease has been reviewed in the AP rat by Tucker (1986), who confirms the observations of other workers that spontaneous bone disease is rare in the rat (Sokoloff, 1967; Burek, 1978; Woodward and Montgomery, 1978). In the

AP rat the sternum, lumbar vertebrae and femur are the bones which have been taken for histological examination in regulatory toxicology studies. After fixation the bones are decalcified by a rapid decalcifying agent and then processed as for other tissues.

Fractures, inflammation and necrosis

Fractures are rare (<1 per cent) and have been found most frequently in the tibia. This low incidence, when compared with other species such as the cat or dog (Hogg, 1948), is considered to be a reflection of the confined, sedentary life of the laboratory rat. Osteomyelitis has occurred as a secondary condition after traumatic injury, such as a fracture, or in neoplastic disease of the bone; it has not been seen as a haematogenous condition. Aseptic bone necrosis is also an uncommon condition in the AP rat (<5 per cent), seen chiefly in the vertebral epiphyses. The pathogenesis of osteonecrosis in the rat is not well defined (Sokoloff and Habermann, 1958), but in other species disturbance of blood supply has been cited as the cause (Dubielzig *et al.*, 1981; Yamasaki and Itakure, 1988).

Chondromucoid degeneration

The most common change in bone is multifocal chondromucoid degeneration. This is seen most frequently at the intercostal junctions of the sternum (Figure 12), but may also occur, rarely, in other articular cartilage. The frequent observation in the sternum may be related to the development of the bone, where the sternal segments (sternebrae) do not fuse and are replaced by bony plates. It is very common (up to 60 per cent) in the sternum of animals in 2 year studies, but does not progress to more severe degenerative conditions.

Osteoporosis

Osteoporosis, an abnormal decrease in bone mass, has been diagnosed in the bones of male AP rats with severe renal disease. In these animals there was marked thinning of the diaphysis of long bones and other changes resembling renal osteitis fibrosa cystica in humans. Apart from the osteoporosis there is marked marrow fibrosis with extensive osteoclast activity (Figure 13) and small cysts in bone and bone marrow. Osteoporosis can be induced in the rat by large doses of retinoic acid, a metabolite of vitamin A (Dhem and Goret-Nicaise, 1984). The practice of some diet manufacturers in adding excessive levels of vitamin A to a diet, to ensure that there is no deficiency, may be considered harmful while the effects of low level excess are not known.

Skeletal growth is virtually continuous in the rat although the maximum growth occurs in the first 4 months of life (Watanabe *et al.*, 1980), and bone mass and calcium homeostasis are regulated by a complex endocrine system; this system is affected by many factors such as age, food intake, body weight,

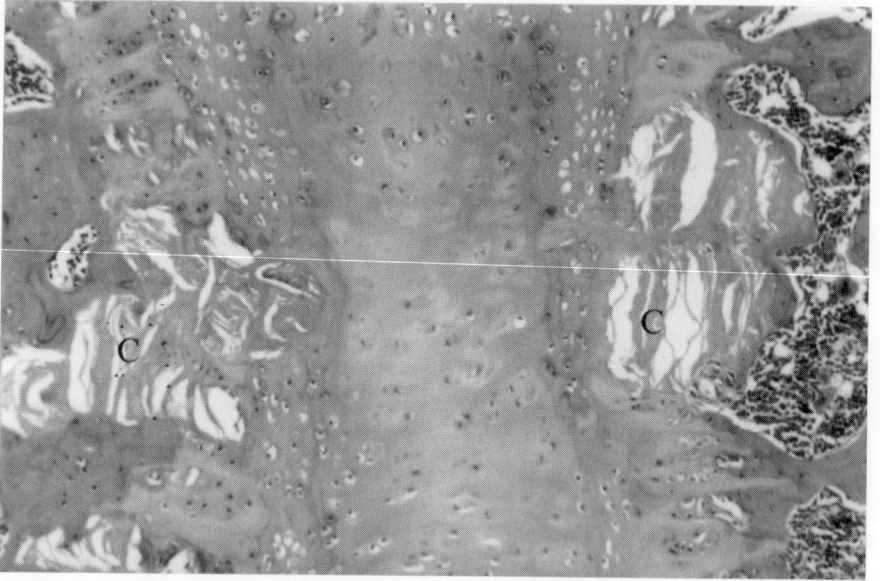

Figure 12 Chondromucoid degeneration (C) at an intercostal junction in the sternum of a 26 month old AP rat

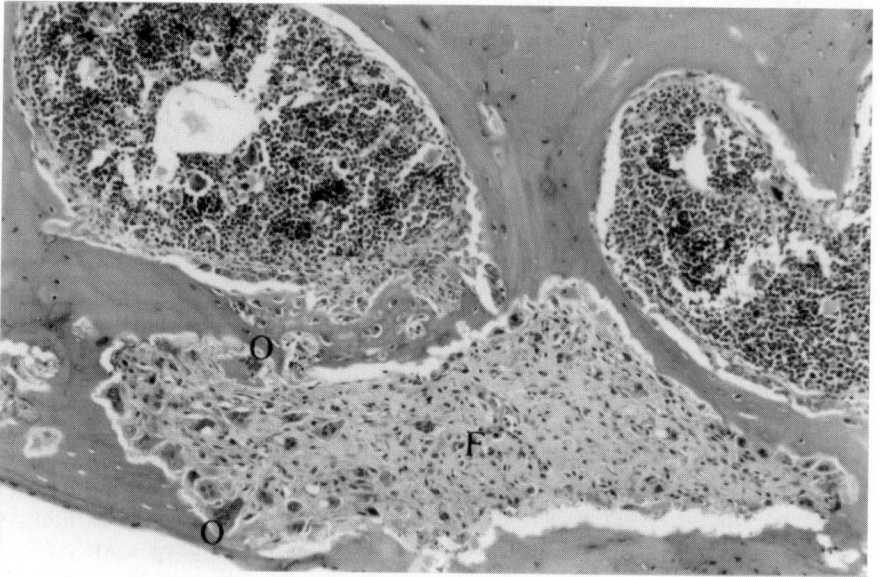

Figure 13 Osteoporosis in the femur of a 22 month old male AP rat showing marrow fibrosis (F) with marked osteoclast (O) activity. ×32,H&E

oestrogen levels and disease (Wade, 1975; Creuss and Hong, 1979; Gray and Wade, 1981; Kalu *et al.*, 1984). Calcium absorption from the intestines is regulated by vitamin D, and calcium levels in the blood are controlled, within a narrow range, by the parathyroid hormone (PTH) and calcitonin (CT). PTH increases renal tubular absorption of calcium and stimulates the renal production of the hormonal form of vitamin D, to increase intestinal absorption of calcium. Spontaneous renal disease in old rats may, therefore, have marked effects on calcium homeostasis, with the end result of decreasing available calcium. Decreased intestinal calcium absorption occurs in old animals, and also bone and kidneys become less responsive to PTH (Kalu *et al.*, 1982), yet serum calcium levels have been shown to remain unchanged, in animals up to 24 months of age, by Smith and Kiebzak (1994). They also showed that the femur increased in length up to 12 months and that calcium levels in the diaphysis of the femur decreased between 12 and 24 months of age. The femurs of female rats had greater calcium levels than males and showed greater density and breaking strength. Whether this is related to the activity of oestrogen, or the absence of renal disease in female rats, has not yet been clarified. Kalu *et al.* (1988) have shown that food restriction in the F344 rat markedly reduced the age-associated increase in PTH, the loss of bone, and the hyperparathyroidism; this supports the contention that it is the decline in renal function which is the cause of bone changes in old male rats.

Hyperostosis

Another change in the bones, particularly in vertebrae, resembles osteitis deformans with irregular cement lines in a mosaic pattern. Hyperostosis, in which there is proliferation of bone in the marrow cavity, is seen occasionally in old rats. The incidence is much lower than that recorded for the F344 rat (Thurman and Bucci, 1994).

3.3.2 Neoplastic Changes

The overall incidence of osteogenic tumours in the database is 0.1 per cent, with the highest incidence in any one study of 2 per cent in males and 1 per cent in females. Approximately half of the 24 2 year studies in the database do not include an osteogenic tumour. Histologically the majority of osteosarcomas were well differentiated with large areas of heavily calcified osteoid; others were admixed with more cellular areas of spindle cells and a few were very pleomorphic with little osteoid and numerous multinucleate giant cells resembling osteoclasts. In some, immature cartilage has been present. This type of histological appearance of osteosarcomas has been described for other strains (Ruben *et al.*, 1986). Table 3.1 shows the types and location of the tumours. It can be seen that the most common site is the vertebrae, followed by the limbs. All of the tumours located in the vertebrae had very extensive metastases in the lungs,

Table 3.1 Incidence of osteosarcomas in the AP rat

| Location of tumour | Number of osteosarcomas[a] | | Number with metastases |
	Male	Female	
Vertebrae	5	0	5
Limbs	3	0	1
Skull	0	2	0
Mandible	1	0	0
Tail	1	1	0
Rib	0	1	0
Not identified[b]	1	1	2

[a] Number of osteosarcomas from a database of 8880 AP rats (including 2800 males and 2500 females in 2 year studies) used in toxicology studies between 1960 and 1994.

[b] Primary site not identified, lung metastases only.

which were considered to be the cause of death, rather than the primary tumour. Metastases were also seen in liver, spleen and abdominal mesentry in some animals. In two animals the primary site was not identified; only lung metastases were found and radiographs of the skeleton did not reveal the site of the primary tumour. The majority of tumours were found in animals over 20 months of age but the youngest was identified in an animal of 18 months of age. No tumours of cartilage have been found and only one periosteal sarcoma in the femur of a male aged 26 months. The gender difference in the incidence of bone tumours is similar to that seen in man (2:1, male:female).

Tumours of bone are rare in all strains of rat (Litvinov and Soloviev, 1973). Burek (1978) recorded only five tumours (0.7 per cent) in his review of 670 rats of the BN/Bi, Wistar WAG/Rij and the hybrid of these two strains. One tumour was found in a 13 month old male but the other four occurred in animals more than 26 months of age. In the F344 the incidence ranges from 0 to 2 per cent (Sass *et al.*, 1975; Maekawa *et al.*, 1983; Stinson, 1990). Tumours of bone are more common in cats and dogs (5 per cent) and the long bones are the most common site. This is also the most common site in humans and bone tumours often arise at the site of previous trauma. The low incidence in laboratory rats may be due to the confined housing which reduces the incidence of traumatic injury to the bones.

3.4 References

ABE, J., FUTJII, Y., KUWAMURA, Y. and HIZAWA, K. (1987) Fiber type differentiation and myosin expression in regenerating rat muscles, *Acta Pathologica Japonica*, **37**, 1537–47.

AL ZUBAIDY, A. J. and MALINOWSKI, W. (1984) Spontaneous pineal body tumours (pinealomas) in Wistar rats: a histological and ultrastructural study, *Laboratory Animals*, **18**, 224–9.

ANSVED, T. and LARSSON, L. (1989) Effects of aging on enzyme-histochemical, morphometrical and contractile properties of the soleus muscle in the rat, *Journal of Neurological Science*, **93**, 105–24.

BUREK, J. D. (1978) In *Pathology of Aging Rats*, pp. 159–60, West Palm Beach, Florida: CRC Press.

CREUSS, R. L. and HONG, K. C. (1979) The effect of long term oestrogen administration in the female rat, *Endocrinology*, **104**, 1188–93.

DHEM, A. and GORET-NICAISE, M. (1984) Effects of retinoic acid on rat bone, *Food and Chemical Toxicology*, **22**, 199–206.

DUBIELZIG, R. R., BIERY, D. N. and BRODEY, R. S. (1981) Bone sarcomas associated with multifocal medullary bone infarction in dogs, *The Journal of the American Veterinary Medical Association*, **179**, 64–8.

EDDINGER, T. J., MOSS, R. L. and CASSENS, R. G. (1985) Fiber number and type composition in extensor digitorum longus, soleus and diaphragm muscles with ageing in Fischer 344 rats, *The Journal of Histochemistry and Cytochemistry*, **33**, 1033–41.

EDDINGER, T. J., CASSENS, R. G. and MOSS, R. L. (1986) Mechanical and histochemical characterization of skeletal muscles from senescent rats, *American Journal of Physiology*, **251**, C421–30.

EVERITT, A. V., SHOREY, C. D. and FICARRA, M. A. (1985) Skeletal muscle ageing in the hind limb of the old male Wistar rat: inhibitory effect of hypophysectomy and food restriction, *Archives of Gerontology and Geriatrics*, **4**, 101–15.

GRAY, J. M. and WADE, G. N. (1981) Food intake, body weight and adiposity in female rats: actions and interactions of progestins and antioestrogens, *American Journal of Physiology*, **40**, E474–E481.

HOGG, A. H. (1948) Osteodystrophic disease in the dog with special reference to rubber jaw (renal osteodystrophy) and its comparison with renal rickets in the human, *Veterinary Record*, **60**, 117–22.

KALU, D. N., HARDIN, R., MURATO, I. *et al.* (1982) Age dependant modulation of parathyroid hormone action, *Age*, **5**, 25–9.

KALU, D. N., HARDIN, R. R., COCKERHAM, R. and YU, B. P. (1984) Aging and dietary modulation of rat skeleton and parathyroid hormone, *Endocrinology*, **115**, 1239–47.

KALU, D. N., MASORO, E. J., YU, B. P., HARDIN, R. R. and HOLLIS, B. W. (1988) Modulation of age-related hyperparathyroidism and senile bone loss in Fischer rats by soy protein and food restriction, *Endocrinology*, **122**, 1847–53.

KRINKE, G., NAYLOR, D. C., SCHMID, S., FRÖHLICH, E. and SCHIDER, K. (1985) The incidence of naturally occurring brain tumours in the laboratory rat, *Journal of Comparative Pathology*, **95**, 175–92.

LAWRENCE, G. M., WALKER, D. G. and TRAYER, I. P. (1986) Histochemical evidence of changes in fuel metabolism induced in red, white and intermediate fibres of streptozotocin-treated rats, *Histochemical Journal*, **18**, 203–12.

LITVINOV, N. N. and SOLOVIEV, J. (1973) Tumours of the bone, in TURUSOV, V. S. (Ed.), *Pathology of Tumours in Laboratory Animals*, Vol. 1, pp. 169–84, Lyon: IARC.

45

MAEKAWA, A., KUROKAWA, Y., TAKAHASHI, M., KOBUBO, T., OGIU, T., ONODERA, H., TANIGAWA, H., OHNO, Y., FURUKAWA, F. and HAYASHI, Y. (1983) Spontaneous tumors in the F344/DuCrj rats, *Gann*, **74**, 365–72.

REZNIK, G., WARD, J. M. and REZNIK-SCHULLER, H. (1980) Ganglioneuromas in the adrenal medulla of F344 rats, *Veterinary Pathology*, **17**, 614–21.

RUBEN, Z., ROHBACHER, E. and MILLER, J. E. (1986) Spontaneous osteogenic sarcoma in the rat, *Journal of Comparative Pathology*, **96**, 89–94.

SASS, B., RABSTEIN, L. S., MADISON, R., NIMS, R. M., PETERS, R. L. and KELLOFF, G. L. (1975) Incidence of spontaneous neoplasms in F344 rats throughout the natural life-span, *Journal of the National Cancer Institute*, 54, 1449–56.

SMALE, G., BENDELE, A. and HORTON, W. E., Jr. (1995) Comparison of age-associated degeneration of articular cartilage in Wistar and Fischer 344 rats, *Laboratory Animal Science*, **45**, 191–4.

SMITH, R and KIEBZAK, G. M. (1994) Effects of aging and exercise on the skeleton, in MOHR, U., DUNGWORTH, D. L. and CAPEN, C. C. (Eds), *Pathobiology of the Aging Rat*, pp. 549–69, Washington: ILSI Press.

SOKOLOFF, L. (1967) Articular and musculoskeletal lesions in rats and mice, in COTCHIN, E. and ROE, F. J. C. (Eds), *Pathology of Laboratory Rats and Mice*, pp. 373–90, Oxford: Blackwell Scientific.

SOKOLOFF, L. and HABERMANN, R. T. (1958) Idiopathic necrosis of bone in small laboratory animals, *Archives of Pathology*, **65**, 323–30.

STINSON, S. F. (1990) Spontaneous tumors in Fischer rats, in STINSON, S. F., SCHULLER, H. M. and REZNIK, G. (Eds), *Atlas of Tumor Pathology of the Fischer Rat*, p. 17, Boca Raton, Florida: CRC Press.

THURMAN, J. D. and BUCCI, T. J. (1994) Hyperostosis in the F344 rat, in MOHR, U., DUNGWORTH, D. L. and CAPEN, C. C. (Eds), *Pathobiology of the Aging Rat*, pp. 565–9, Washington: ILSI Press.

TUCKER, M. J. (1986) A survey of bone disease in the Alpk/AP rat, *Journal of Comparative Pathology*, **96**, 197–203.

WADE, G. N. (1975) Some effects of ovarian hormones on food intake and body weight in female rats, *Journal of Comparative and Physiological Psychology*, **88**, 183–93.

WATANABE, M., TANAKA, H., KOIZUMI, H., TANIMOTO, Y., TORII, R. and YANAGITA, T. (1980) General toxicity studies of tamoxifen in mice and rats, *Jitchuken Zenrinsho Kenkyuho*, 6, 1–36.

WOODWARD, J. C. and MONTGOMERY, C. A. (1978) The musculoskeletal system, in BENIRSCHKE, K., GARNER, E. M. and JONES, T. C. (Eds), *Pathology of Laboratory Animals*, pp. 663–880, New York: Springer-Verlag.

YAMASAKI, K. and ITAKURE, C. (1988) Aseptic necrosis of bone in ICR mice, *Laboratory Animals*, **22**, 51–3.

4

The Digestive System

4.1 Oral Cavity

4.1.1 Non-neoplastic Changes

Teeth

The teeth and oral cavity are inspected routinely in toxicology studies with the AP rat but, apart from the tongue, only tissues which appear abnormal are sampled for histological examination. The most common change seen in the teeth is malocclusion, which occurs in 1 per cent of animals, and broken incisors which are even less frequent. Both conditions interfere with feeding and usually necessitate sacrifice of the affected animals. Dietary fibres which penetrate the gingiva can give rise to gingivitis and periodontitis, and occasionally abscesses. Such changes have been seen rarely, except in a few studies where the animals were fed powdered diet rather than the usual pelleted diet. Similar changes were reported by Robinson (1985). The dietary pellets fed to AP rats are firm, but not hard, and probably reduce dental problems by providing appropriate wearing of the occlusal surfaces and preventing overgrowth. Very hard diets can produce a variety of problems due to abrasive loss of enamel and dentin (Bossman *et al.*, 1981). Histological examination of teeth has been done in the AP rat in several studies which were specifically designed to examine teeth. These were short term studies and only a few minor changes were seen, including some deposition of secondary dentin and pulpal stones. In old AP rats the incisors may become discoloured due to porphyrin deposits in the enamel.

4.1.2 Neoplastic Changes

The only neoplastic changes observed in the teeth were two ameloblastomas (Figures 14 and 15) in males aged 14 and 16 months, respectively. All of the neoplasms seen in the oral mucosa were squamous carcinomas (Figure 16) which were found only in those studies where animals were fed powdered rather than pelleted diet. One of these studies was the life-span study where all of the oral tumours were found in rats older than 26 months; the incidence was similar in both sexes and reached an overall incidence of 4 per cent. As previously described these studies also had a high incidence of inflammatory lesions from fibre penetration of the oral mucosa. It seems likely that the tumours are related to long-standing infections caused by the fibre penetration. It has been suggested by Buckley *et al.* (1980) that the AP rats which develop this disease have an inherited tendency to develop squamous carcinomas. They drew this conclusion after they obtained a breeding nucleus of AP rats to set up a colony in their own laboratory at the MRC Radiobiology Unit. Inbreeding of their colony resulted in a 50 per cent incidence of oral squamous carcinomas.

Tumours of the oral cavity are rare in other strains of rat. Burek (1978) reported four squamous carcinomas, all in female rats, and two had lymph node metastases. Stinson and Kovatch (1990) reported a low incidence of oral tumours in the F344 rat, and dental tumours have been observed infrequently in other strains (Lewis *et al.*, 1980; Fitzgerald, 1987; Ernst and Mohr, 1991).

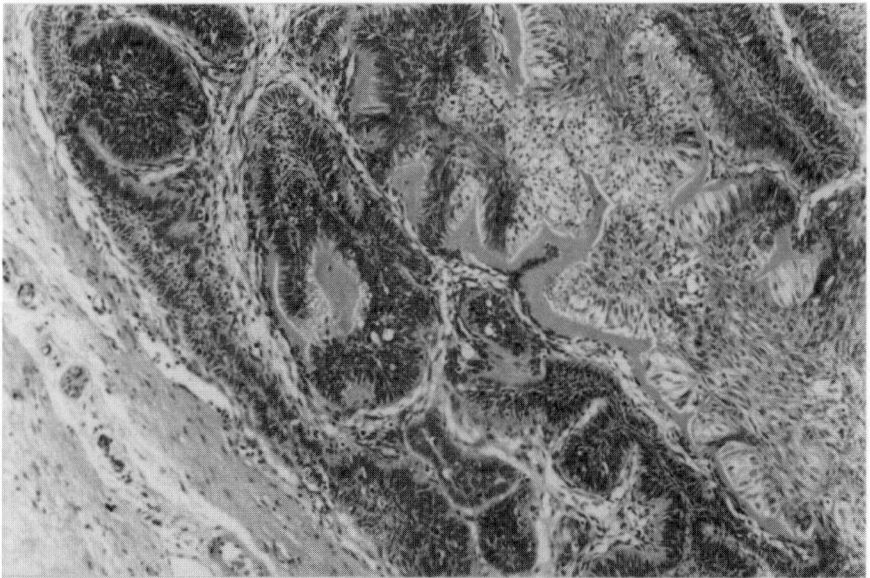

Figure 14 Ameloblastoma arising in the mandible of a male AP rat aged 14 months. × 32, H&E

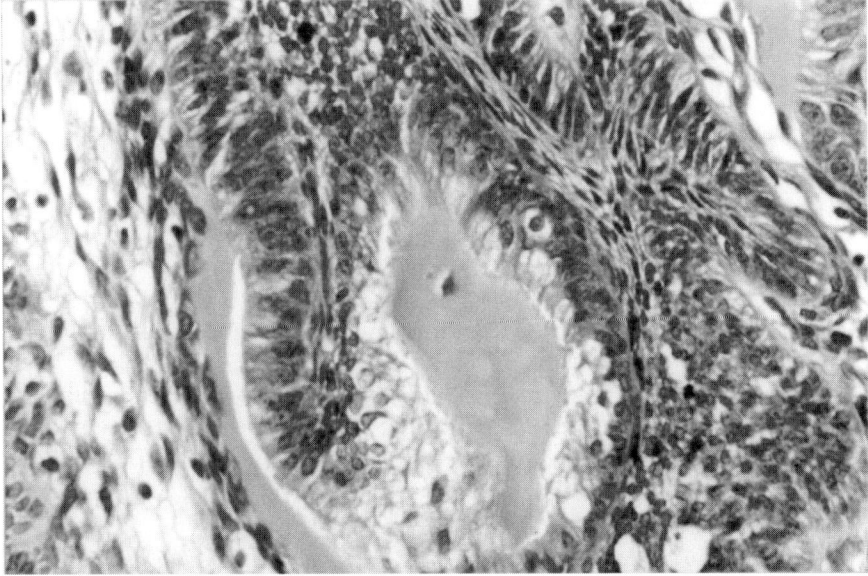

Figure 15 Ameloblastoma. ×128, H&E

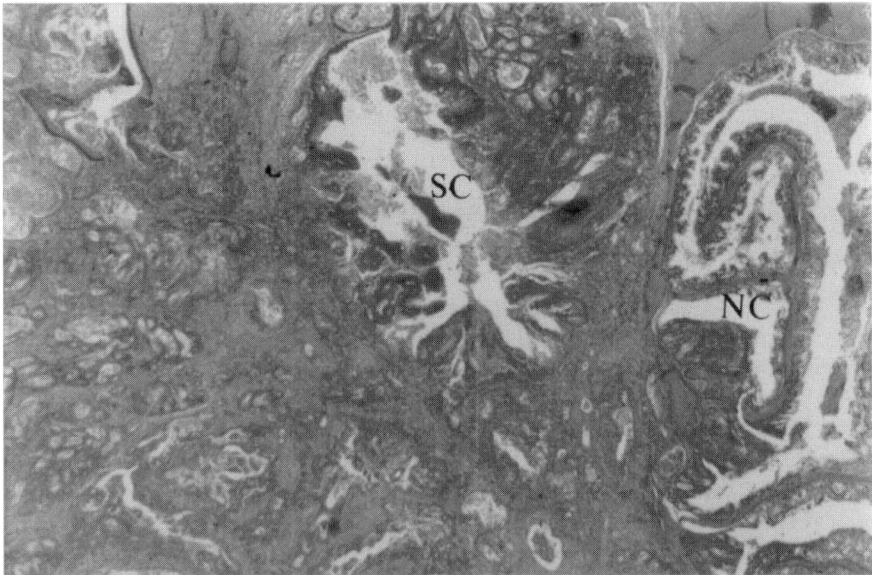

Figure 16 Invasive squamous carcinoma (SC) of the oral mucosa adjacent to the nasal cavities (NC). × 8, H&E

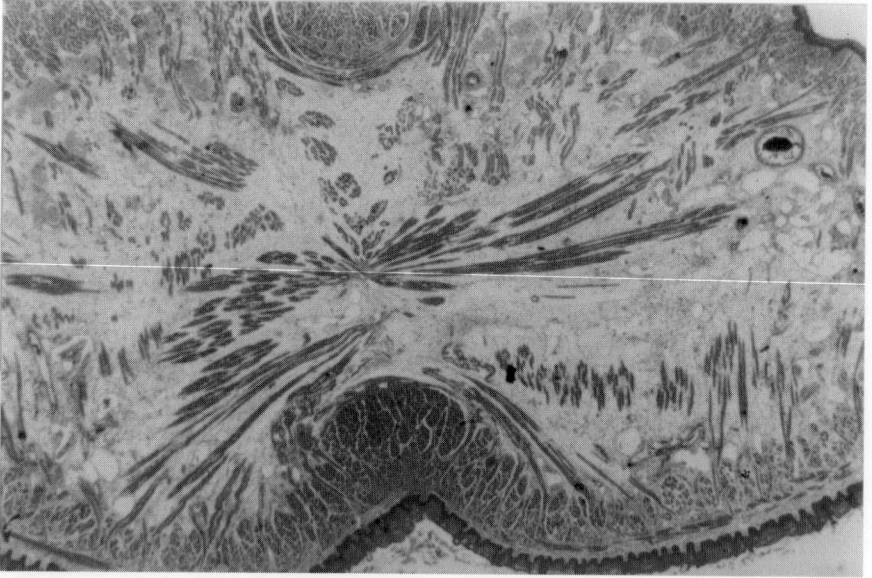

Figure 17 Tongue showing glossitis with oedema and inflammation of muscle. ×8, H&E

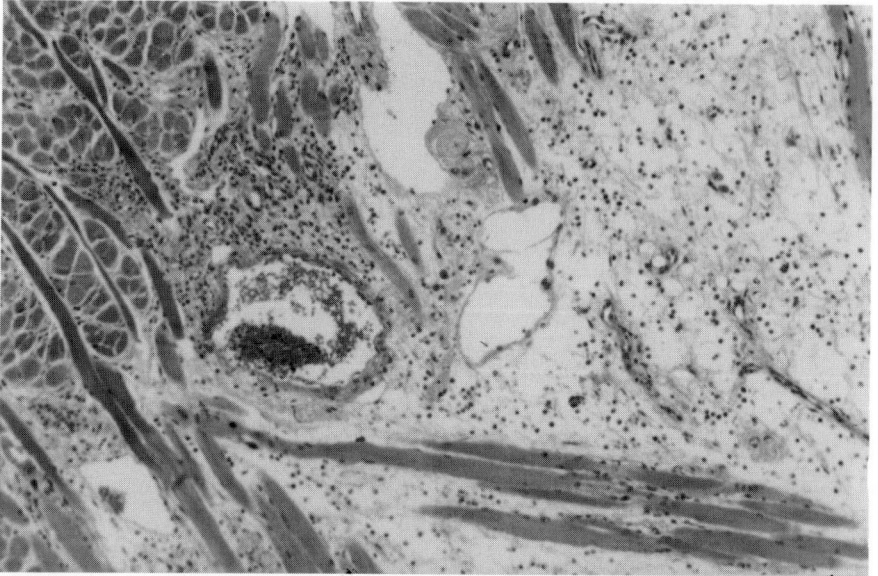

Figure 18 Glossitis with oedema and a mild inflammatory cell infiltrate. ×32, H&E

4.2 Tongue

4.2.1 *Non-neoplastic Changes*

Inflammation

The tongue is examined routinely in toxicology studies in the AP rat but has shown few lesions. They include glossitis (Figures 17 and 18), arteritis and some variations in the thickness of the epithelial layer. Sialoadenitis of the lingual salivary glands at the base of the tongue is another rare observation.

4.2.2 *Neoplastic Changes*

Neoplastic lesions of the tongue in the AP rat are confined to two squamous papillomas. This low incidence is common to most rat strains (Burek, 1978; Stinson and Kovatch, 1990), although Kociba and Keyes (1985) reported an incidence of 3.2 per cent of squamous cell tumours in male SD rats.

4.3 Salivary Glands

4.3.1 *Non-neoplastic Changes*

Inflammation

The three salivary glands of the rat are the mucous secreting sub-lingual gland, the serous secreting parotid gland and the mixed secreting submandibular (submaxillary) gland. All three are examined in regulatory toxicology studies in AP rats. Mononuclear cell infiltrates, of varying size, are common in the parotid and, to a lesser extent, in the submandibular glands. The latter gland also shows considerable nuclear variability in the cells of the serous secreting glands (Figure 19).

The only condition in the salivary glands which produces overt clinical signs is sialoadenitis caused by the sialodacyoadenitis rat corona virus. This has been observed in a long term study in AP rats. Routine monitoring of serum, from sentinel animals, in the particular study where the disease occurred did not identify the virus. In this 2 year study, which was infected in 1975, almost all of the 390 rats, in all groups, developed the infection during the eighth month of the study. Swelling of the neck and signs of respiratory distress were almost universal clinical signs. Several of the rats died, but the majority recovered and at the end of the 2 year study the salivary glands showed little evidence of any sequelae. The rats which died showed the characteristic features of the disease, with marked oedema, inflammation and hyperplasia of duct epithelium (Figure 20) in the parotid and submandibular glands. Rapid recovery from the infection has also been seen by other workers (Carthew and Slinger, 1981). The source

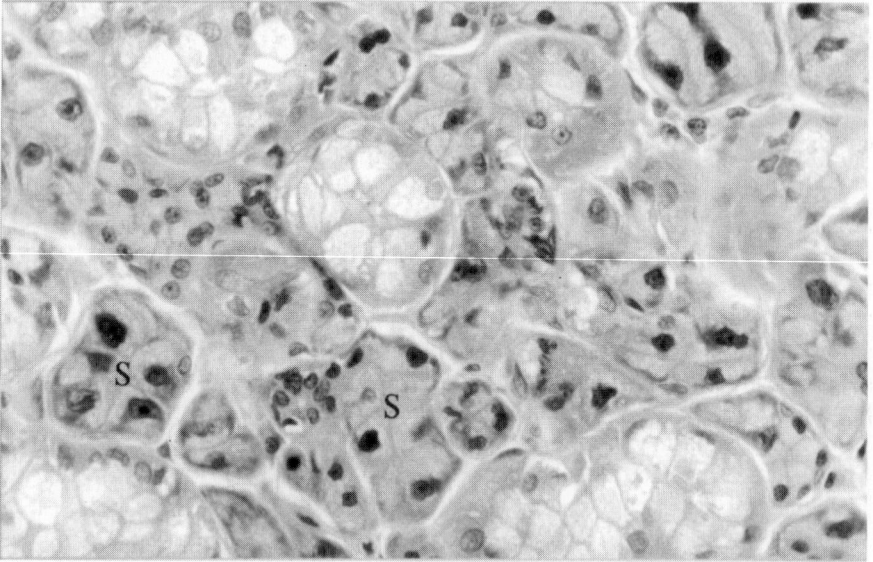

Figure 19 Submandibular salivary gland showing nuclear variation in the cells of serous glands (S). ×128, H&E

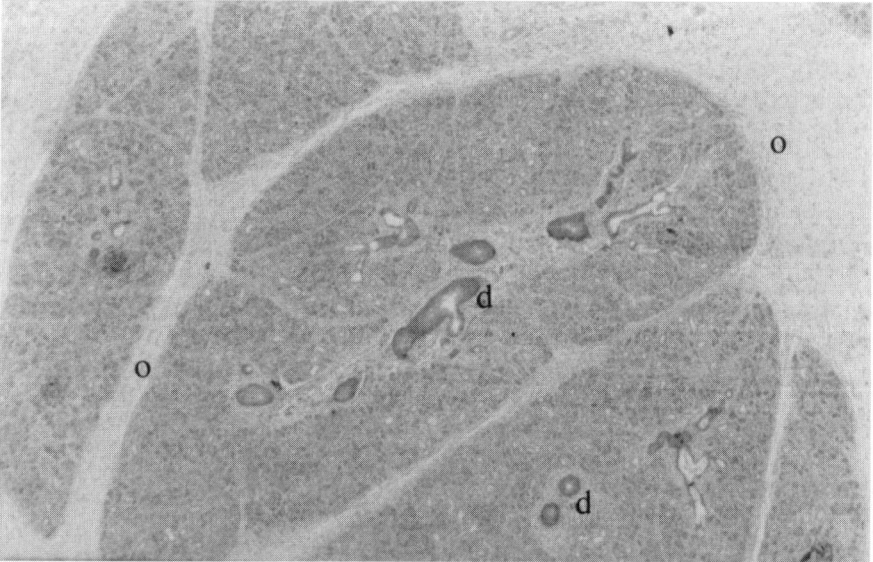

Figure 20 Sialoadenitis in the submandibular salivary gland of a male AP rat showing interlobular oedema (o) and marked duct epithelial hyperplasia (d). ×8, H&E

of the infection is unknown as the animals were kept in a room where they had no contact with any other rodents at any time. As the animals did not develop overt signs of disease until the eighth month it seems unlikely that they had acquired the virus in the breeding colony.

Cytomegaly

The parotid gland frequently shows areas of acinar cytomegalic change (Figure 21). The cells in these areas have an extensive cytoplasm and enlarged hypochromatic nuclei, but are not associated with inflammatory changes, or increased mitotic activity; nor are there any specific clinical signs of virus infection. Chiu and Chen (1986) consider that they are hypertrophic changes and they do not represent a pre-neoplastic change (Dawe, 1979). This is supported by the incidence in the AP rat where cytomegaly is quite common but tumours of the salivary gland are rare.

Functional changes

Functional changes in the salivary glands are known to be related to the secretion of sex hormones (Liu *et al.*, 1969; Dean and Hiramoto, 1984) thus, as animals age and levels of these hormones decline, the salivary glands show a reduced secretory activity with a decrease in the size of acini. Removal of the submandibular gland has been linked by Boyer *et al.* (1990) to a decrease in plasma levels of luteinising hormone. Decreased food consumption or reduced

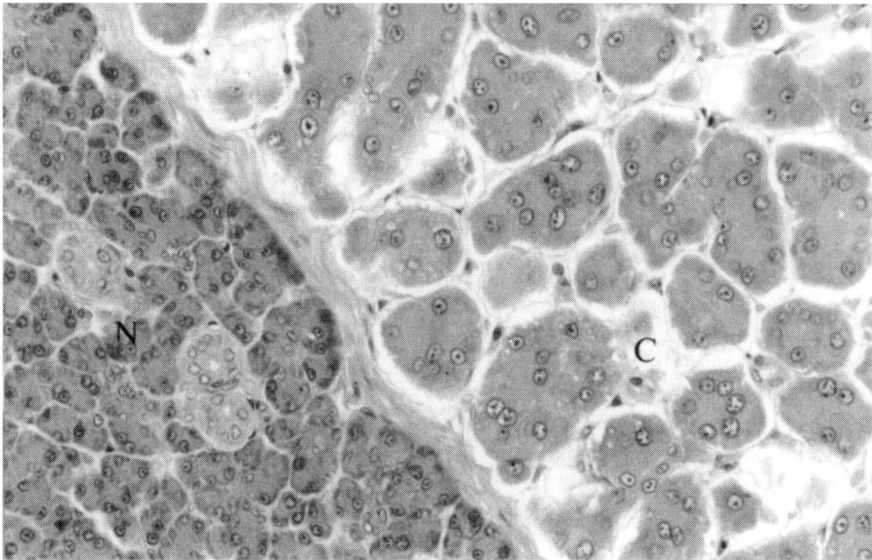

Figure 21 Cytomegalic change in the parotid salivary gland of a male AP rat showing normal acini (N) and enlarged pale staining cytomegalic acini (C). ×128, H&E

protein intake also reduces the weight of the glands and causes shrinkage of secretory acini (Boyd *et al.*, 1970).

4.3.2 Neoplastic Changes

Neoplasia is rare in all of the salivary glands of AP rats, and no tumours have been seen in the sub-lingual gland. The maximum incidence in any study is 2 per cent, but a zero incidence is most common. The salivary gland tumours found in the AP rat include both epithelial and, less frequently, mesenchymal tumours. The numbers observed are shown in Table 4.1 and include adenomas and adenocarcinomas of acinar cells in both parotid and submandibular glands and a duct adenoma in the parotid gland. These were all well-differentiated tumours with cells in acinar patterns closely resembling the acinar patterns of the gland from which they are derived. The benign tumours were all found in animals killed at the end of 2 year studies (aged 26 months) and the malignant tumours were found in animals between 20 and 26 months of age. Tumours of the salivary gland are rare in other strains (Elwell and Leininger, 1990). Burek (1978) recorded three tumours in 670 rats, while the highest incidence in the 284 bioassays in the F344 rats in the National Toxicology Program (NTP) was 0.2 per cent (Stinson and Kovatch, 1990). Secondary tumours are rare in the salivary glands but soft tissue tumours arising in adjacent connective tissue have been observed to invade the glands, and infiltration by leukaemias and lymphomas has been seen in a few animals.

Table 4.1 Incidence of tumours of salivary glands in AP rats

Tumour	Number observed [a]	
	Males	Females
Acinar adenoma parotid	3	6
Acinar adenoma submandibular	1	0
Adenocarcinoma parotid	2	0
Adenocarcinoma submandibular	0	1
Duct adenoma parotid	1	0
Fibrosarcoma submandibular	3	0
Malignant fibrohistiocytic- sarcoma submandibular	1	0

[a] Histological types of tumour observed in the submandibular and parotid salivary glands of the AP rat from a database of 8880 animals (including 2800 males and 2500 females in 2 year studies) used in toxicology studies between 1960 and 1992. The maximum incidence in any of 24 oncogenicity studies was 2% acinar adenomas of the parotid in female rats.

4.4 Oesophagus

4.4.1 *Non-neoplastic Changes*

Congenital anomalies

The whole oesophagus is examined in toxicology studies in the AP rat; the area between the mediastinum and larynx is taken unopened for histological examination, and the remainder of the oesophagus is opened and inspected. Spontaneous histological lesions are rare but a congenital abnormality seen infrequently (<0.1 per cent) is distension of the oesophagus with hypertrophy and degeneration of the muscle wall. This condition is fatal, usually within the first 6 months of life. Dilatation of the oesophagus has been reported by Maita *et al.* (1986) in F344 rats, with an incidence of 1.3 per cent in males only, and in females a 1 per cent incidence of hyperkeratosis of the oesophagus; both conditions were only found in animals killed at 109 weeks of age. A condition termed megaoesophagus has been described by Harkness and Ferguson (1979). This condition is characterised by an enlarged oesophagus and degeneration of muscle fibres and ganglion cells. Oesophageal impaction due to food or bedding in the oesophagus was described in BHE rats by Ruben *et al.* (1983).

Inflammation

In the AP rat, trauma from catheterisation is not uncommon, particularly in long term studies where dosing was by gavage. The most common change is a mild abrasion of the lining epithelium with varying degrees of inflammation and hyperkeratosis. Perforation of the oesophagus has been observed, occasionally, with escape of fluid or food into the localised area which may produce a severe inflammation which extends to cause pleuritis and/or pericarditis. Clinical signs of respiratory distress precede death in the more severe cases.

4.4.2 *Neoplastic Changes*

Neoplasms are rare in all strains (Pozharisski, 1973a), including the AP rat. The database includes only two tumours, both from animals in the life-span study. One was a well-differentiated squamous papilloma in a male aged 27 months (Figure 22), the other a locally invasive squamous carcinoma in a female killed in a moribund condition at 33 months. A single papilloma was found in 2000 Wistar rats by Bomhard *et al.* (1986), a squamous carcinoma in one of 786 males of the Rochester-strain Wistar rat (Crain, 1958), an undifferentiated sarcoma in one out of 1945 Osborne-Mendel rats (Goodman *et al.*, 1980), and none in the F344 rats of the NTP bioassays (Stinson, 1990).

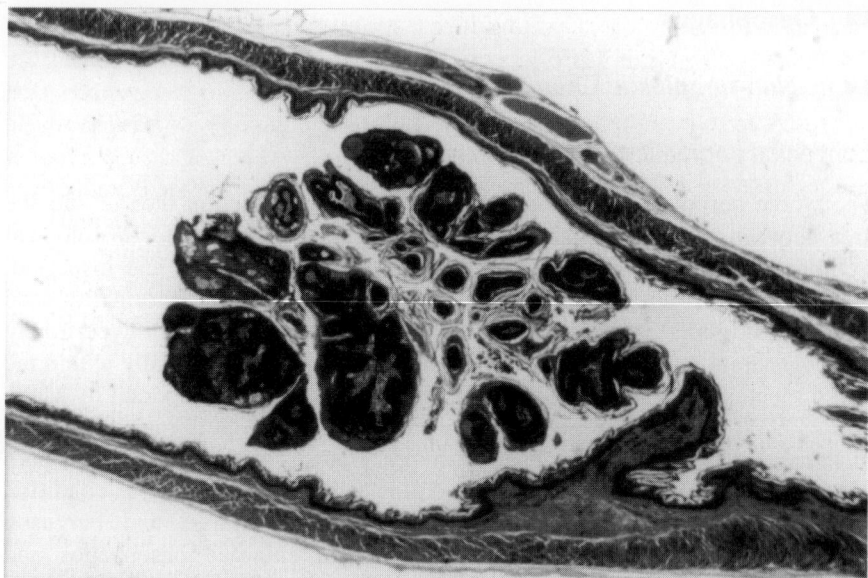

Figure 22 Squamous papilloma of the oesophagus in a 27month old male AP rat. The section does not pass through the stalk. ×8, H&E

4.5 Stomach

4.5.1 *Non-neoplastic Changes*

The procedure for examination of the stomach in the AP rat is to open it along the greater curvature and wash the contents away by gently shaking the stomach in physiological saline. The routine sections include one which passes through the proximal area (forestomach), the limiting ridge and the fundus; a second section is taken through the pyloric antrum. The only congenital abnormality seen very occasionally in the AP rat is a squamous cyst in the submucosa of the fundus or forestomach.

Inflammation

Inflammation is the most common finding (maximum incidence of 5 per cent) and includes erosions and ulceration, chiefly in the forestomach, and may be single or multiple. These changes are most common in older animals (>18 months). Histological changes range from the least severe, where there is loss of the squamous epithelium associated with small numbers of inflammatory cells, to large ulcers extending down to the muscle wall with widespread inflammation of the crater and extensive oedema of the submucosa. In the fundus, haemorrhagic erosions of the superficial mucosa are the more common form of inflammation; penetrating ulcers are exceedingly rare. The cause of forestomach ulceration has not been established, although several factors are known to influence the development of ulcers, including starvation, age, infections and stress. In the AP rat they are most common in males with severe renal disease or females with large pituitary tumours. This would support either stress or starvation as a possible cause since food intake is greatly reduced in both renal disease and in the presence of large pituitary tumours. Acanthosis, hyperkeratosis and parakeratosis of the forestomach epithelium occurs infrequently (<1 per cent) in the absence of ulcers or inflammation, and these conditions do not appear to have any clinical significance as the animals do not loose body weight or condition. Greaves and Faccini (1984) recorded similar changes in the forestomach which were so florid that it was difficult to make a clear distinction between hyperplasia and neoplasia. Vitamin A deficiency is known to produce such changes in the rat (Klein-Szanto *et al.*, 1982) but this is not likely to be the cause in AP rats as their diet contains the recommended levels of the vitamin.

Scattered dilatation of crypt glands in the fundus is a common finding in rats in all studies longer than 1 month; they increase with age and show no sex difference in incidence. Similar changes have been recorded in F344 rats by Brown and Hardisty (1990) and Iwata *et al.* (1991). They may be associated with mucosal atrophy in very old (>2 years old) AP rats. In this condition there is also some loss of glands, chief and parietal cells, with replacement by fibrous connective tissue. Mineralisation of the stomach occurs in animals with severe renal disease, with calcification of the muscle wall throughout the stomach, and also in the mucosal glands of the fundus and pylorus.

4.5.2 Neoplastic Changes

Epithelial tumours

Neoplastic disease in the stomach of the AP rat is rare (<1 per cent) although there is a range of different histological types as shown in Table 4.2. The squamous papilloma is the most common epithelial tumour, and in the AP rat they were all single tumours clearly visible at necropsy. Histologically, papillomas were similar to the example in Figure 23, with connective tissue strands lined by a squamous epithelium showing hyperkeratosis. The single malignant squamous carcinoma, in a 31 month old female, was poorly different-iated with pleomorphic cells, only occasional pearl formation, and a high mitotic rate. It had invaded a large area of the fundus and extended into the peritoneum. It is of interest that, although the spontaneous incidence of tumours in the forestomach is so low, induction of forestomach tumours has been found with a wide range of chemicals and naturally occurring substances (Nagayo, 1973). This would suggest that the rat forestomach is sensitive to carcinogens and the low spontaneous incidence in the rat indicates little or no exposure to dietary

Table 4.2 Incidence of tumours of the stomach in the AP rat

Histological type	Total number observed[a]
Squamous papilloma forestomach	3
Squamous carcinoma forestomach	1
Adenocarcinoma fundus	1
Fibrosarcoma	5
Leiomyosarcoma	2
Malignant fibrohistiocyticsarcoma	1
Malignant lymphoma	1

[a] Numbers of different histological types of tumour observed in the stomach of the AP rat from a database of 8880 control rats (including 2800 males and 2500 females in 2 year studies) used in toxicology studies between 1960 and 1994. The highest overall percentage incidence, or the maximum percentage incidence of any one type, observed in 24 oncogenicity studies was 1%.

carcinogens. An adenocarcinoma of the fundus has been seen in one animal in a group dosed with a compound; it was not thought to be related to treatment.

Mesenchymal tumours

Mesenchymal tumours of the stomach tend to be small and confined to the wall of the stomach. They were all found in animals more than 22 months of

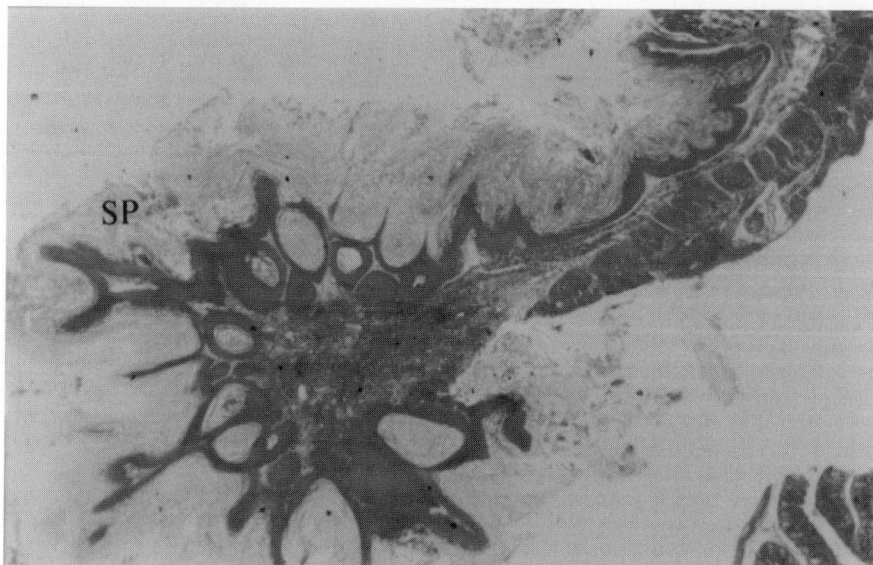

Figure 23   Squamous papilloma (SP) in the forestomach of a male AP rat. ×8, H&E

age. Masson's trichrome was used to distinguish the tumours derived from fibrous and muscle tissue, as they both had a histological appearance of interlacing spindle cells which made it difficult to distinguish them. The fibrohistiocytic sarcoma and lymphoma (which was lymphoblastic) both had the characteristic appearance of the tumours at other sites. The lymphoma was found only in the stomach; the more usual sites for lymphomas – lymph nodes, spleen, bone marrow and thymus – were not involved. Stomach tumours are rare in all strains (Rowlatt, 1967; Fukushima and Ito, 1985; Takahashi and Hasegawa, 1990), including the F344 (Goodman *et al.*, 1979; Stinson and Kovatch, 1990; Tatematsu and Imaida, 1990a), SD rats (Imai and Yoshimura, 1988), and the Osborne-Mendel rat (Goodman *et al.*, 1980).

4.6 Intestines

4.6.1 *Non-neoplastic Changes*

Congenital anomalies/nematodes

Samples of duodenum, jejunum, ileum, caecum and colon are taken, unopened, for histological examination; the remainder of the intestines are opened and inspected. Ectopic pancreas in the duodenal submucosa is a rare congenital abnormality (<0.1 per cent). In the early years of the colony, when the AP rat was fed Powder 'O' diet, nematodes were occasionally found in the intestines and these were identified as cat nematodes. Contamination of the diet was presumed to occur at the mill where the diet was formulated, as rodenticides were not permitted and cats were used to keep down the rodent population. Due to changes in the source and preparation of the rodent diet, nematodes have not been seen in the intestines for approximately 20 years.

Inflammation

Spontaneous inflammatory conditions are uncommon in the AP rat as in most SPF strains, where the burden of pathogenic micro-organisms is minimal. Chronic enteritis and duodenal ulceration are most common as conditions secondary to neoplastic changes (maximum incidence of 4 per cent). A similar low incidence is reported for F344 rats by Maekawa (1994). Arteritis of submucosal blood vessels is seen when there is widespread vascular disease, and mineralisation of muscle wall and mucosa occurs in hyperparathyroid disease secondary to renal disease and is most common in the colon (Figure 24). Very rare conditions which have only been seen once or twice include diverticula, intersusception and herniation through diaphragm or the umbilicus. Peyer's patches in the ileum are not usually prominent but can be detected at

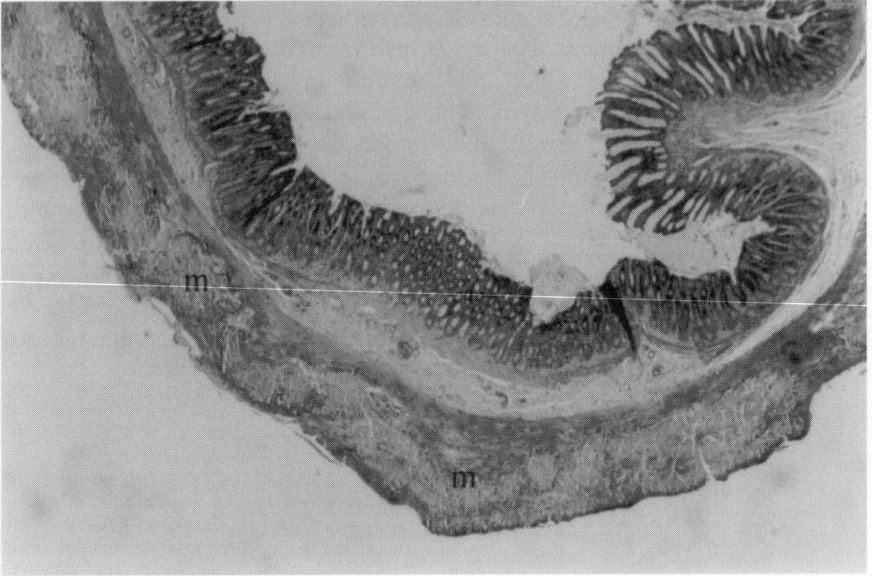

Figure 24 Mineralisation (m) in the muscle wall of the colon in a male AP rat with severe renal disease, parathyroid hyperplasia and widespread dystrophic calcification. ×32, H&E

necropsy when they are enlarged, and histologically they show enlarged germinal centres with increased numbers of lymphoblasts. This may occur in the absence of any other change in the ileum or related to neoplastic or inflammatory conditions.

4.6.2 Neoplastic Changes

Neoplastic diseases of the intestines are rare, with mesenchymal tumours as common as epithelial tumours. The maximum incidence of intestinal tumours in any one study was 3.4 per cent, but was only 2 per cent for any one type. The histological types observed are shown in Table 4.3.

Epithelial tumours

The most common epithelial tumour is a well-differentiated adenomatous polyp with a distinct pedicle (Figure 25). The majority have been found in the colon but a few have been seen in the ileum and jejunum, but not the duodenum. These were mostly incidental findings at necropsy in animals killed at termination of 2 year studies; a few were large enough to cause obstruction, loss of body weight and a deterioration in the general health of the animal which necessitated unscheduled necropsy. Intestinal tumours have not been

Table 4.3 Incidence of tumours of intestines in the AP rat

Histological type	Total number observed[a]
Adenomatous polyp duodenum	1
Adenomatous polyp jejunum	2
Adenomatous polyp ileum	1
Adenomatous polyp colon	5
Tubulo-papillary adenocarcinoma duodenum	1
Mucus secreting adenocarcinoma colon	1
Mucus secreting adenocarcinoma jejunum	1
Scirrhous adenocarcinoma jejunum	1
Scirrhous adenocarcinoma colon	1
Leiomyoma duodenum	2
Leiomyosarcoma duodenum	1
Leiomyosarcoma jejunum	1
Leiomyosarcoma ileum	1
Fibrosarcoma duodenum	2
Fibrosarcoma jejunum	3
Fibrosarcoma ileum	3
Fibrosarcoma colon	1
Lymphoblastic lymphosarcoma ileum	3

[a] Numbers of different histological types of intestinal tumour observed in the AP rat from a database of 8880 control animals (including 2800 males and 2500 females in 2 year studies) used in toxicology studies between 1960 and 1994. The highest overall percentage incidence observed in any of 24 oncogenicity studies was 3.4%, and the highest percentage incidence of any one type was 2%.

seen in animals less than 18 months of age. Osseous metaplasia has been a feature of several polyps.

Mesenchymal tumours

Fibrosarcoma and leiomyosarcoma are equally common and have been observed in all areas of the intestines, and again only occur in older animals. Adenocarcinomas include tubular, papillary, mucus-secreting and scirrhous tumours. The two mucus-secreting tumours showed cells with intracellular accumulations of mucus to give the characteristic 'signet ring' appearance. An extensive connective tissue stroma surrounding clumps of undifferentiated epithelial cells was a histological feature of the scirrhous adenocarcinomas. The fibrosarcomas and leiomyosarcomas were all well-differentiated tumours apparently arising in the muscle wall and invading the mucosa. The three primary lymphomas of Peyer's patches were lymphoblastic (two) or lymphocytic (one) lymphomas and had extended out of the greatly enlarged Peyer's patches to infiltrate the wall of the adjacent ileum.

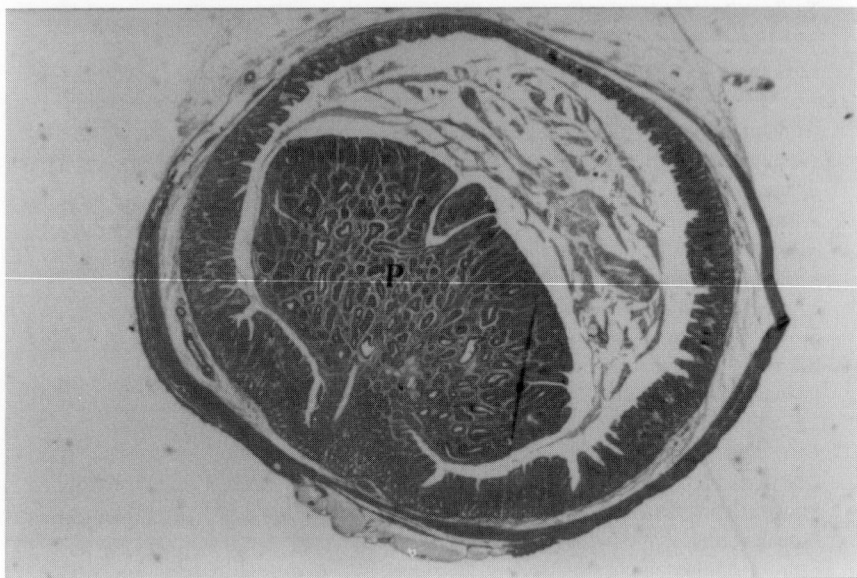

Figure 25   Well-differentiated adenomatous polyp (P) of the colon in a male AP rat. ×8, H&E

The type and incidence of intestinal tumours in the AP rat are similar to those cited for other strains. Pozharisski (1973b) cited the 30 spontaneous intestinal tumours in the literature to that time. More recently, similar incidences have been recorded for the SD rat (Imai and Yoshimura, 1988), F344 (Maekawa *et al.*, 1983a; Maita *et al.*, 1986; Tatematsu and Imaida, 1990b), Osborne-Mendel (Goodman *et al.*, 1980) and Wistar (Maekawa *et al.*, 1983b).

4.7 Liver

4.7.1 Non-neoplastic Changes

The rat liver has four major lobes, and in the AP rat samples for histological examination are taken from the median and left lateral lobes and fixed by immersion except for studies specifically designed to examine liver morphology or function. In these studies perfusion fixation is considered essential (Roberts *et al.*, 1990). After sampling for histological examination, the remainder of the liver is sectioned and the cut surface examined for any abnormality. The liver is the most extensively studied organ in the rat, since it is the most common site for pathological change induced by xenobiotics. The AP rat, however, does not have any spontaneous disease in the liver which produces clinical signs or significant mortality, although there is an extensive range of histological changes.

Liver weights

Liver weights are important in toxicology since many xenobiotics affect weight, and age-related changes in liver weight in control animals reflect the metabolic activity of the liver. Table 4.4 gives the absolute liver weight and relative weight (liver as per cent of body weight) at different time points. There is little difference in the absolute weights of the liver between 32 and 108 weeks, indicating that significant liver growth ceases at some point between 12 and 34 weeks. Relative liver weight declines from 12 to 58 weeks as body weight increases but rises again at 108 weeks when spontaneous diseases have developed. Similar trends in weights were reported for Sprague-Dawley COBS rats (Irisarri and Hollander, 1994).

Clear cell change

Clear cell change can be demonstrated to be due to the presence of glycogen; it is usually seen, in formalin fixed H&E stained sections, as a lacy cytoplasm with spaces where the glycogen has been removed during processing (Figure 26). In young AP rats the hepatocytes show plentiful glycogen at all times, although the amount increases at night during feeding; it diminishes in centrilobular cells during the day when the animals are asleep and fasting. It can be depleted rapidly from all cells when inanition occurs in neoplastic and renal disease. This is one example of the effect of diet upon hepatic histology. It has been shown by Berlin *et al.* (1982) that changes in the hepatocyte content of glycogen, lipids and the endoplasmic reticulum are related to the amount of carbohydrate, protein and fat in the diet. High protein diets have been shown to be associated with active gluconeogenesis in the liver (Didier *et al.*, 1985) and some inbred strains of rat accumulate glycogen in the liver as a result of a deficiency of hepatic phosphorylase kinase (Haynes *et al.*, 1983). Changes in lighting periods have also been shown to alter liver glycogen (Bhattacharya, 1983). There cannot, therefore, be a standard histological appearance for a 'normal' hepatocyte as the histological appearance is affected by several different environmental factors. Any disease which causes a significant

Table 4.4 Liver weights at different time points

Age (weeks)	Number/ sex	Mean liver weight (g)		Mean relative liver weight (% body weight)	
		Males	Females	Males	Females
12	10	17.0	10.2	5.0	4.5
34	20	22.0	12.3	3.7	3.8
58	25	22.2	13.0	3.0	3.4
108	100	21.6	16.0	3.5	3.7

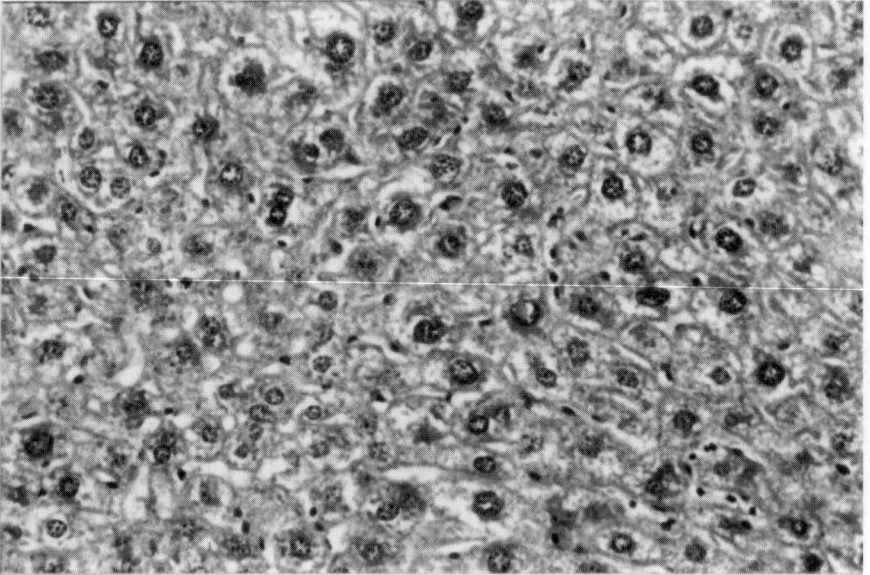

Figure 26 Hepatocyte clear cell change demonstrated by spaces in the cell cytoplasm where glycogen has been removed during histological processing. ×128, H&E

reduction in food intake will quickly produce a loss of glycogen from hepatocytes.

Apart from the presence of glycogen, the most consistent feature of the hepatocyte in H&E stained sections is the basophilic stippling which corresponds to the rough endoplasmic reticulum (RER). Schmucker (1990) has shown that there is a reduction in RER between 6 and 20 months of age; subsequently the level increases again, which is thought to correspond to an increased production of albumin by the liver. This increased production is possibly a compensatory response to the albuminurea which occurs in old rats.

Steatosis

Fat vacuolation (steatosis) may occur as multiple small lipid-containing vacuoles (microvesicular) or single large vacuoles (macrovesicular) in the cytoplasm of the hepatocyte. The distribution of fat vacuolation can be focal, diffuse or zonal; the most common variety is periportal fat vacuolation. This occurs in females which have severe inanition due to the presence of large pituitary tumours. Diffuse fat vacuolation (Figure 27) is rare. Zimmerman (1978) has suggested that steatosis is due to failure of removal of lipid from the liver cell rather than an increased entry into the cell. Greaves and Faccini (1984) attribute spontaneous fat vacuolation to nutritional, metabolic or hormonal effects. The occurrence of fat vacuolation in the AP rat is rare before 12 months of age, but an incidence of 30 per cent in males and 50 per cent in

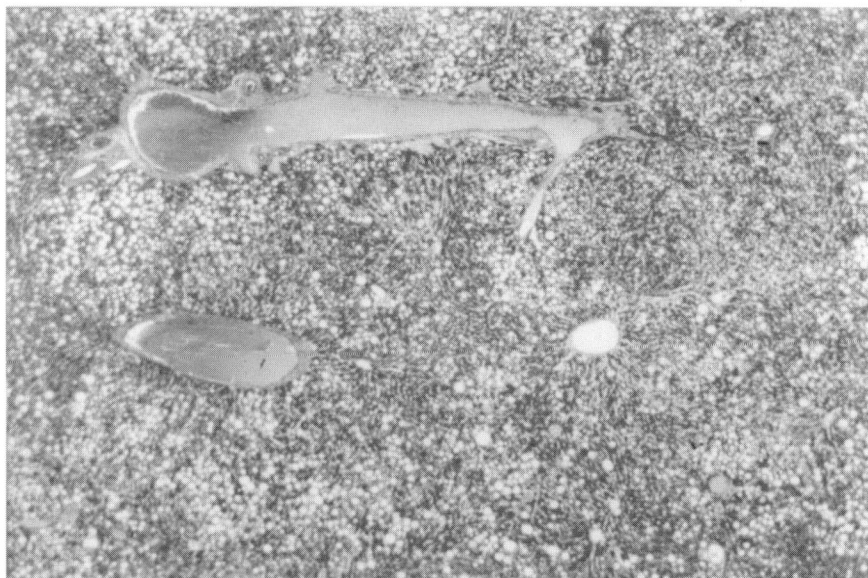

Figure 27 Diffuse hepatocyte fat vacuolation. ×32, H&E

females has been seen in 2 year studies, although the severity in most animals is minimal to mild. The other fat storing cells of the liver, the Ito cells, may also accumulate lipid in otherwise normal livers, or in hypervitaminosis A (Wake, 1974).

Necrosis

Necrosis can be single cell, zonal, focal or diffuse. The most common type in the AP rat is focal necrosis, where clearly defined single or multiple foci of necrotic cells are present; the size of the foci varies from only a few cells to almost a lobule in size and may have associated acute or chronic inflammation, although this is not always present. It produces a minimal effect on liver function, with only slight elevations of AST and ALT. Focal necrosis increases up to 12 months of age, where incidence levels of 20 per cent have been observed, but thereafter the incidence declines. There is a 2:1 difference in incidence in males:females. The cause of this necrosis is not known; it is not associated with infections such as Tyzzer's disease or enteritis. Kountouras *et al.* (1984) demonstrated that focal necrosis can occur following bile duct obstruction, but there is no evidence of obstruction in the AP rat.

Centrilobular necrosis is rare in young animals but occurs in 3–4 per cent of males and females in 2 year studies, as a terminal condition in animals dying from renal or neoplastic diseases. Infarction with total necrosis has been observed in both anterior and posterior portions of the caudate lobe. In the early stages there is very extensive necrosis and haemorrhage (Figure 28) which

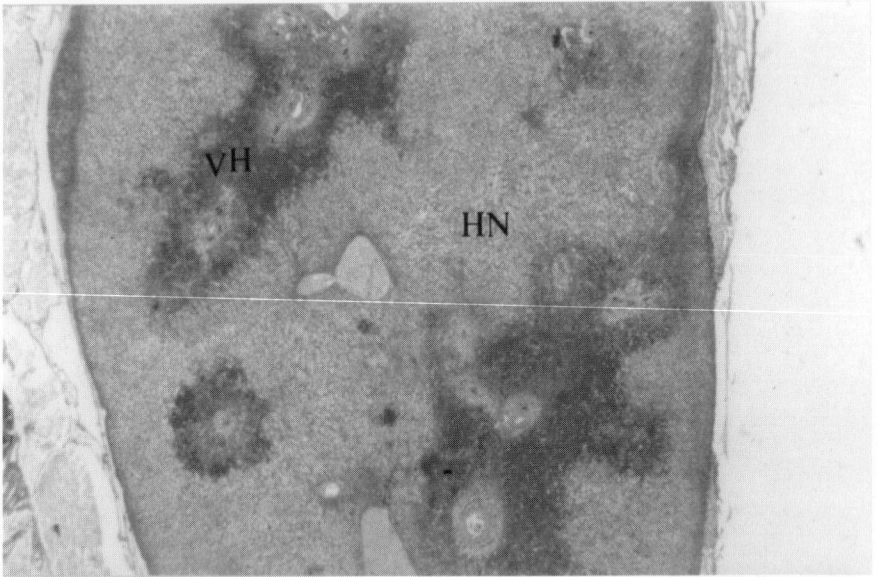

Figure 28   Infarct of the caudate lobe of the liver with large areas of necrosis (HN) and a few remaining viable cells around blood vessels (VH). ×8, H&E

resolves into a fibrotic mass with extensive pigmentation and only a few viable liver cells (Figure 29). The fibrotic process may become detached from the liver and be found, at later necropsy, as a yellow body free floating in the upper abdomen. This is a rare condition, and the cause is likely to be occlusion of a major artery caused by torsion of the lobe (Davidsohn *et al.*, 1963).

Single cell necrosis (apoptosis) may be seen in the liver of young animals but declines with age as do the numbers of mitotic figures which may be seen frequently in the liver of young animals (Figure 30). Apoptosis in the liver may be increased in animals with various diseases in organs other than the liver.

Bile duct proliferation

The most common condition seen in the AP rat liver is proliferation of bile ducts, which occurs in animals from 12 months, and at 2 years is present in 50 to 80 per cent of males and 40 to 64 per cent of females. This condition has become much more frequent with time. In the late 1960s the incidence was 8 to 10 per cent. The severity of the condition varies and may consist of only a few foci of proliferating ducts (Figure 31) to large numbers of foci, which may become confluent. Within the foci are a number of bile ducts, some dilated, lined by a flat to cuboidal epithelium, and associated with fibrosis and infiltration of the stroma by plasma cells and lymphocytes (Figure 32). Similar changes have been noted in other strains but the incidence is reported to be low (Burek, 1978; Stewart *et al.*, 1980). The aetiology of this change is not known

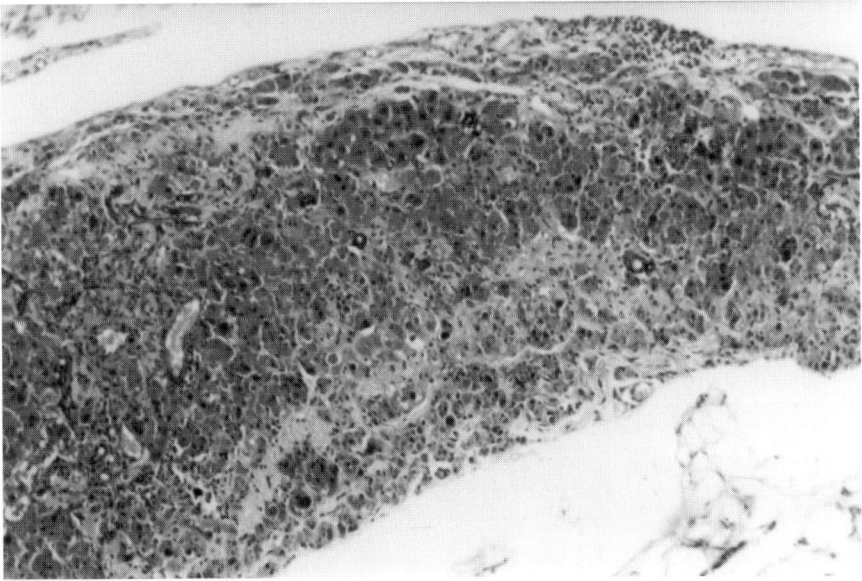

Figure 29 Infarcted caudate lobe of liver with a few viable cells and areas of fibrosis and pigmentation. ×32, H&E

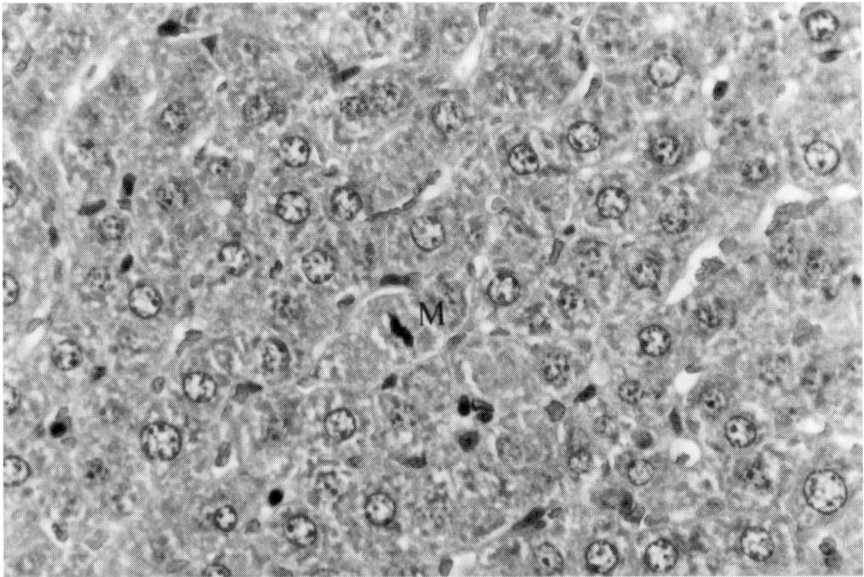

Figure 30 Hepatocyte showing mitotic figure (M). ×128, H&E

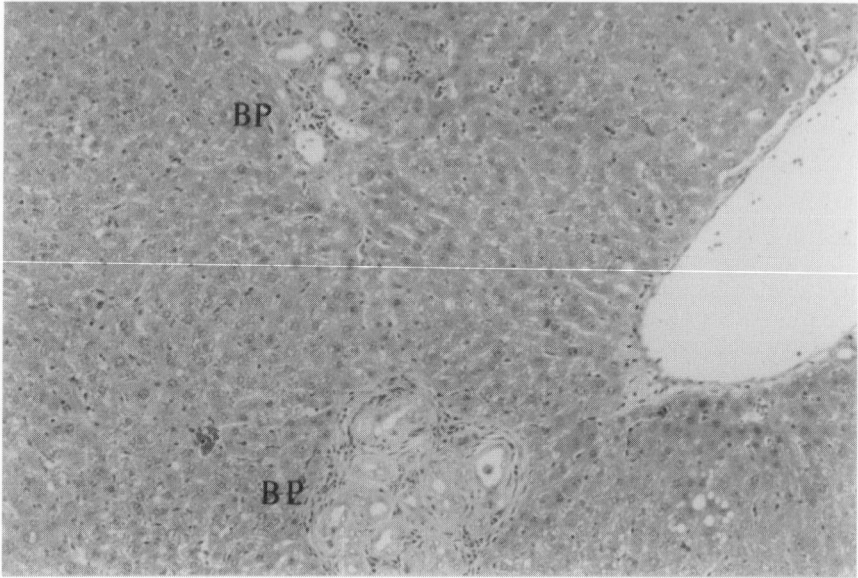

Figure 31 Bile duct proliferation. × 8, H&E

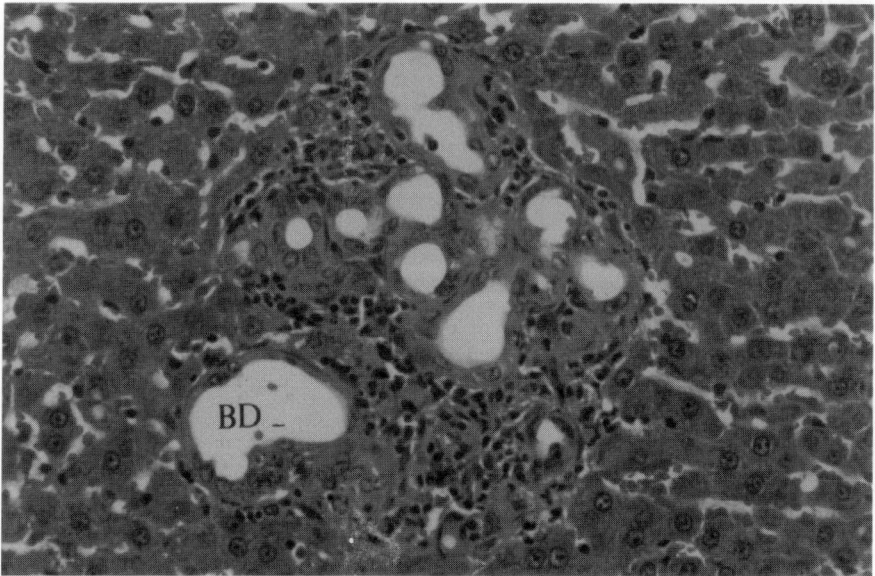

Figure 32 Foci of bile duct proliferation (BD) in the liver. ×32, H&E

but it is reduced in AP rats subjected to a 20 per cent dietary restriction (unpublished data) and also in restricted F344 rats (Yu *et al.*, 1982). The increased incidence in the AP rat over the last 30 years may be related to the increasing size and increased food intake of the animal. Large single bile-filled cysts are rare (<1 per cent) but were very common in the Wistar rats investigated by Burek (1978).

Inflammation

Inflammatory cell infiltrates, sometimes granulomatous, are a common finding in the liver of rats of all ages, and may be associated with necrotic cells. They do not appear to have any important effect upon liver function. The origin of this change in the AP rat is not known but rat parvovirus, which usually only affects the neonatal rat, can be reactivated in adult animals by immunosuppression (Jacoby *et al.*, 1979).

Pigmentation

Pigmentation of hepatocytes with ceroid, haemosiderin or lipofuscin occurs to some extent in almost all AP rats over 18 months of age. They are sometimes referred to as the pigments of degeneration and are thought to be related to changes in metabolic activity within the cell (Knook, 1982).

Peliosis and spongiosis hepatis

Peliosis hepatis is a rare change in the AP rat. It is composed of blood-filled spaces of varying size, usually without an endothelial lining. It has been described by Boorman and Hollander (1973) in another Wistar strain. Lee (1983) suggested that it is related to disruption of the sinusoidal endothelium and it can be produced in rats by virus infections (Bergs and Scotti, 1967). Spongiosis hepatis is a more common condition, occurring in up to 10 per cent of animals at 2 years, but it is rare before that time. It differs from peliosis in that the cavities are usually devoid of blood and contain a pink mucopolysaccharide material (Figure 33) and are composed of fibroblast-like cells and fat cells (Bannasch *et al.*, 1981).

Amyloid has not been observed in the liver of the AP rat, and cirrhosis of the liver, with gross bridging fibrosis, has been seen in only two male rats.

Altered foci

Altered foci of hepatocytes – clear cell, basophilic and eosinophilic – are uncommon and do not occur before 18 months. They increase with age to a maximum of 10 per cent in males, and 6 per cent in females, with clear cell foci the most common. They are not increased significantly by the administration of genotoxic or non-genotoxic carcinogens. The incidence of foci varies

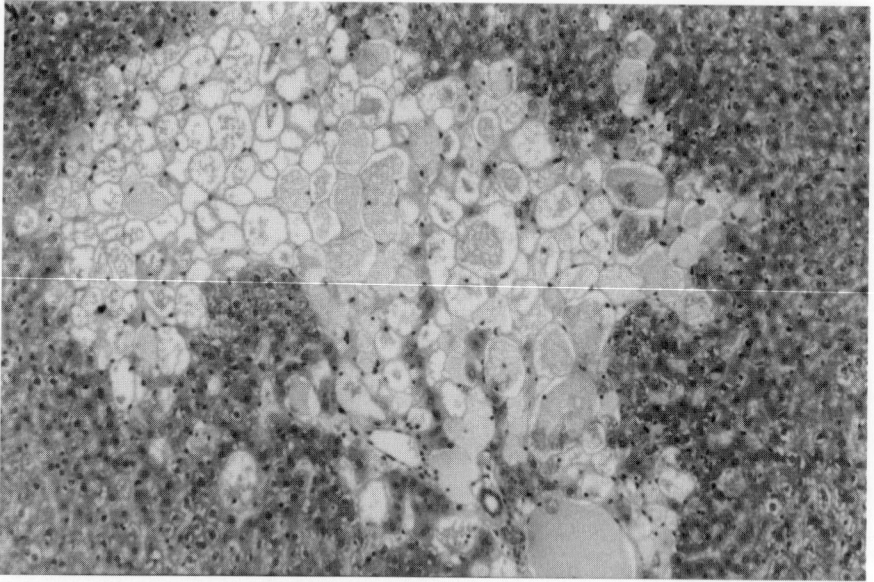

Figure 33 Spongiosis hepatis showing cavities filled with pale staining mucopolysaccharide material. ×80, H&E

considerably between strains. Burek (1978) reported less than 5 per cent in the BN/Bi strain but much higher incidences in the Wistar WAG/Rij strain and the F_1 hybrid of these strains, with almost twice as many foci in females (87 per cent). Ward (1981) reported that in the F344 rat basophilic foci were most common.

Hyperplasia

Hyperplastic nodules are diagnosed by size (less than a lobule) and absence of compressed adjacent cells. They are also rare in the AP rat liver, with a maximum 10 per cent in males and 2 per cent in females.

4.7.2 *Neoplastic Changes*

Hepatocellular tumours

Liver tumours are rare in most strains of rat although they are readily induced by a wide range of xenobiotics. The histological types observed in the AP rat are shown in Table 4.5. There has been no increase in the incidence over the 40 year period the AP strain has been used, with the highest incidence in a 2 year study completed in 1980, and all tumours were found in animals more than 22 months old. The most common type observed is the hepatocellular adenoma,

Table 4.5   Incidence of tumours of liver in the AP rat

Histological type	% Incidence[a]	
	Males	Females
Hepatocellular adenoma	3.3	4.6
Hepatocellular carcinoma	2.0	2.0
Cholangioma	1.5	0
Cholangiocarcinoma	0	1.5
Angioma	2.2	0
Angiosarcoma	0	1.5

[a] Incidence of liver tumours in the AP rat from a database of 8880 control animals (including 2800 males and 2500 females in 2 year studies) used in toxicology studies between 1960 and 1994. The highest incidence of all histological types of liver tumours combined was 4.6%.

which is a solitary, small tumour of relatively normal cells surrounded by compressed cells; there are no portal tracts, bile ducts or centrilobular veins within the tumour. The malignant hepatocellular tumours (a total of three have been observed) show cellular pleomorphism and abnormal trabeculae (Figure 34) and frequent mitotic figures, but are still recognisable as liver cell tumours. They were locally invasive and one had metastatic deposits in the lung.

Bile duct tumours

Only two cholangiomas and one cholangiocarcinoma have been found in untreated AP rats. Both showed clear duct-like structures lined by a flat to cuboidal epithelium; the benign tumours had a fine fibrous capsule and the malignant tumour was less well differentiated and locally invasive. Angiomas were distinguished from peliosis by the presence of endothelium, and the one angiosarcoma observed showed extensive local invasion and areas of undifferentiated endothelial cells.

Metastatic tumours

Secondary (metastatic) tumours of the liver are not common but infiltration by leukaemic cells does occur. Lymphoblastic leukaemias have a distinct infiltration of periportal sinusoids, but myeloid and monocytic leukaemias and lymphomas tend to have a generalised sinusoidal infiltration. Metastases of intestinal and Harderian gland tumours have also been observed.

This low incidence of liver tumours is also reported for the F344, SD and other strains (Schauer and Kunze, 1976; Burek, 1978; Goodman *et al.*, 1980;

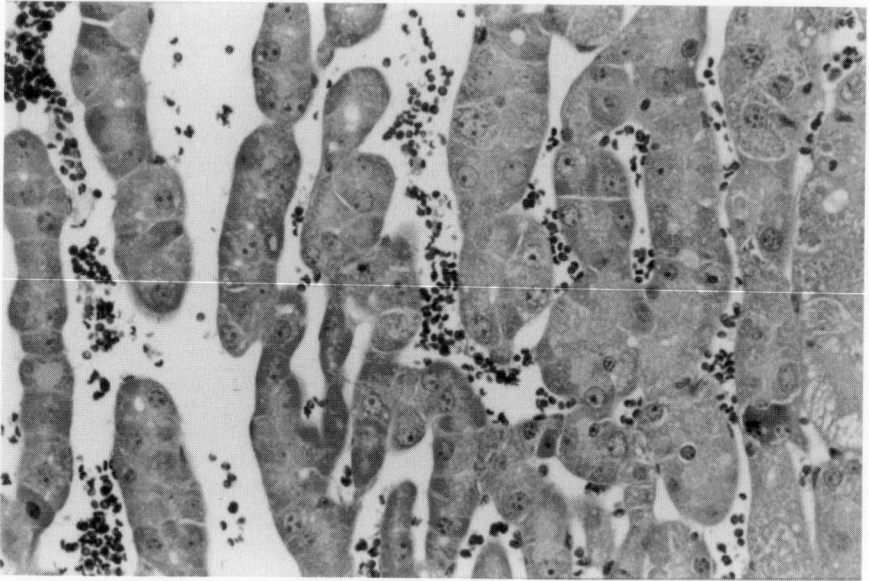

Figure 34   Hepatocellular carcinoma showing abnormal trabecular patterns. ×128, H&E

Maronpot, 1990). Liver tumours have been induced in the AP rat by genotoxic compounds, 2-acetylaminofluorene (unpublished data), quinoxaline 1,4 dioxide (Tucker, 1975) and the non-genotoxic hypolipidaemic fibrate methyl clofenapate (Tucker and Orton, 1995).

4.8   Exocrine Pancreas

4.8.1   *Non-neoplastic Changes*

Atrophy

Atrophy is the most common disease of the exocrine pancreas. Kendry and Roe (1969) consider that loss of acinar cell basophilia is the earliest change. This is followed by mononuclear cell infiltrates and fibrosis of the stroma, and then increasing loss of acinar tissue and replacement by foci of duct-like or cystic areas and adipose tissue (Figures 35 and 36). Atrophy is more common in animals over 18 months of age and reaches a maximum incidence of 10 per cent in males and 2 per cent in females. The incidence varies with strain, with levels of 3.8 to 5.9 per cent in the F344 (Coleman *et al.*, 1977; Goodman *et al.*, 1979), 2 per cent in the Osborne-Mendel (Goodman *et al.*, 1980) and 46 per cent in the males of the BN/Bi/Wistar Rij hybrid (Burek, 1978).

72

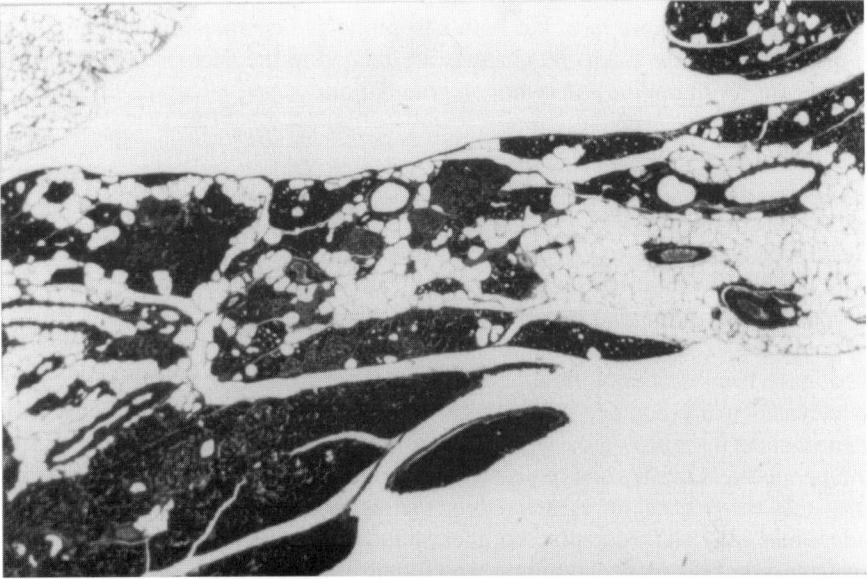

Figure 35 Atrophy of pancreatic exocrine tissue with replacement by adipose tissue. ×80, H&E

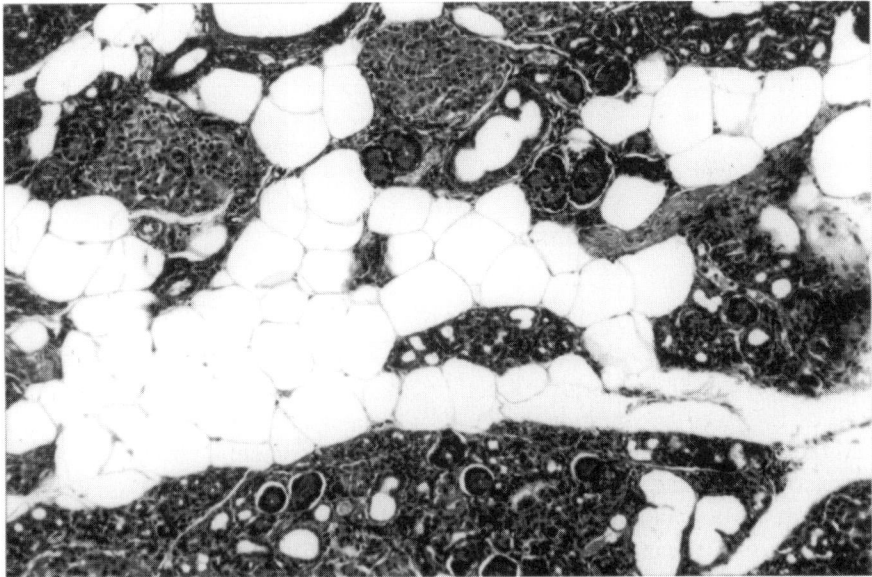

Figure 36 Higher power view of Figure 35 showing pancreatic exocrine atrophy. ×80, H&E

Altered foci

Foci of altered cells are rare, but both basophilic and eosinophilic foci are seen in older animals; the tinctorial changes are related to the decrease (basophilic) or increase (eosinophilic) of zymogen granulation.

4.8.2 Neoplastic Changes

Hyperplastic nodules and adenomas of exocrine cells are not uncommon in male AP rats, with a maximum tumour incidence of 11 per cent (females 1.4 per cent). They have also become more frequent in the last 10 years as shown in Table 4.6. The earliest adenoma seen was in a male aged 15 months, but most have been incidental findings at terminal necropsy. They may be detected as elevated white nodules when large, but many of the adenomas were only identified during histological examination. The distinction between hyperplasia and neoplasia is mostly one of size as the tumours are so well differentiated that only the absence of ducts and islets distinguish them from normal tissue. Adenomas may, infrequently, be encapsulated (Figure 37). A single, locally invasive exocrine adenocarcinoma was found in a male aged 22 months and a duct carcinoma in a female aged 21 months. Exocrine tumours are rare in the rat. Roe and Roberts (1973) recorded a 1 per cent incidence in Chester Beatty Wistar rats, and Berg (1967) a 2–3 per cent incidence in SD rats. Burek (1978)

Table 4.6 Incidence of tumours of exocrine pancreas in the AP rat

Year	Incidence of tumours of exocrine pancreas[a]	
	Male	Female
1960	0	0
1965	0	0
1971	2.0	0
1975	1.4	1.4
1982	11.2	1.3
1985	5.0	1.3
1989	7.4	1.4

[a] Incidence of exocrine pancreatic tumours in the AP rat from a database of 8880 control rats (including 2800 males and 2500 females) used in toxicity studies between 1960 and 1994. A zero incidence was recorded in at least 10 of 24 oncogenicity studies completed before 1980.

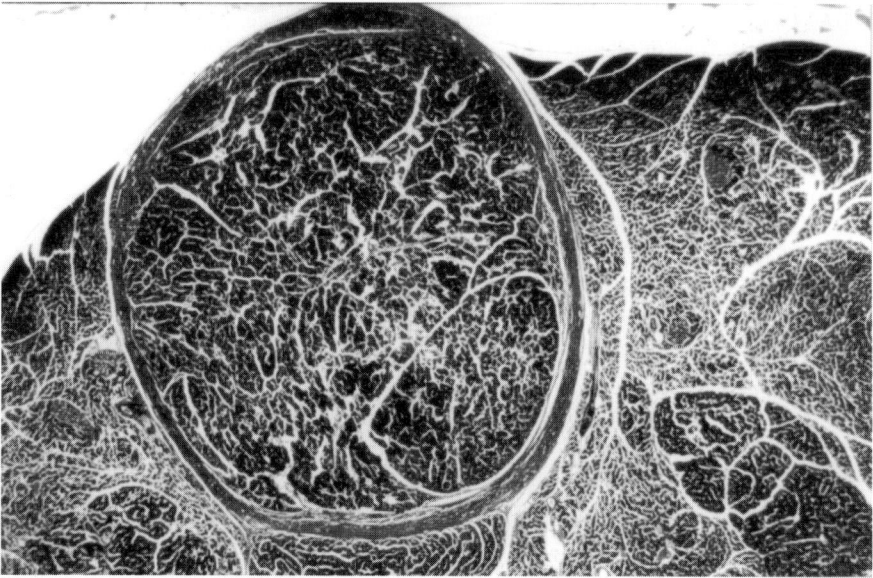

Figure 37 Well-differentiated exocrine adenoma pancreas with fibrous capsule. ×8, H&E

recorded two adenomas in males aged 36 and 40 months while Boorman and Eustis (1990) found the incidence in F344 rats, where the pancreas was sampled, to be 14 per cent in males and 1 per cent in females. In another study where the whole pancreas was examined the incidence rose to 37 per cent in males and 5 per cent in females, but these animals were receiving corn oil as a vehicle. Pancreatic exocrine tumours have been induced in the AP rat by the non-genotoxic hypolipidaemic drug methylclofenapate (Tucker and Orton, 1995). The histological appearance of the tumours was similar to that of the spontaneous tumours. Secondary infiltration by leukaemias and lymphomas occurs infrequently.

4.9 References

BANNASCH, P., BLOCH, M. and ZERBAN, H. (1981) Spongiosis hepatis, Specific changes of the perisinusoidal liver cells induced by N-nitrosomorpholine, *Laboratory Investigation*, **44**, 252–64.

BERG, B. N. (1967) Longevity studies in the rat, in COTCHIN, E. and ROE, F. J. C. (Eds), *Pathology of Laboratory Rats and Mice*, pp. 101–2, Oxford: Blackwell Scientific.

BERGS, V. V. and SCOTTI, T. M. (1967) Virus induced peliosis hepatitis in rats, *Science*, **158**, 377–8.

BERLIN, J., CASTRO, C. E., BAILEY, F. and SEVALI, J. S. (1982) Adaption of rat parenchymal hepatocyte to nutritional variation: Quantitation by stereology, *Nutrition Research*, **2**, 51–63.

BHATTACHARYA, R. D. (1983) Heterogeneity in circadian phase shifting of some liver variables in altered light–dark cycle, *Cellular and Molecular Biology*, **29**, 483–7.

BOMHARD, E., KARBE, E. and LOESER, E. (1986) Spontaneous tumours in 2000 Wistar TNO/W.70 rats in two-year carcinogenicity studies. *Journal of Environmental Pathology, Toxicology and Oncology*, **7**, 35–52.

BOORMAN, G. A. and EUSTIS, S. L. (1990) Neoplasms of the exocrine pancreas, in STINSON, S. F., SCHULLER, H. M. and REZNIK, G. (Eds), *Atlas of Tumor Pathology of the Fischer Rat*, pp. 222–5, Boca Raton, Florida: CRC Press.

BOORMAN, G. A. and HOLLANDER, C. F. (1973) Spontaneous lesions in the female WAG/Rij (Wistar rat) rat, *Journal of Gerontology*, **28**, 152–61.

BOSSMAN, K., DEERBERG, F., PREUSS, V. and REHM, S. (1981) Dental and periodontal alterations in aging Han:WIST rats, *Zeitschrift für Versuchstierkunde*, **25**, 305–11.

BOYD, E. M., CEHN, C. P. and MUIS, L. F. (1970) Resistance to starvation in albino rats fed from weaning on diets containing 0 to 81 per cent of protein as casein, *Growth*, **34**, 99–112.

BOYER, R., ESCOLA, R., BLUET-PAJOT, M. T. and ARANCIBIA, S. (1990) Ablation of submandibular salivary glands in rats provokes a decrease in plasma luteinizing hormone levels correlated with morphological changes in Leydig cells, *Archives of Oral Biology*, **35**(8), 661–6.

BROWN, H. R. and HARDISTY, J. F. (1990) Oral cavity, esophagus and stomach, in BOORMAN, G. A., EUSTIS, S. L., ELWELL, M. R. and MONTGOMERY, C. A. (Eds), *Pathology of the Fischer Rat*, pp. 9–30, San Diego: Academic Press.

BUCKLEY, P., HULSE, E. V. and KEMP, B. M. (1980) An inbred strain of rats with a high incidence of squamous-cell carcinomas of the mouth, *British Journal of Cancer*, **41**, 295–301.

BUREK, D. (1978) in *Pathology of Aging Rats*, pp. 75, Boca Raton, Florida: CRC Press.

CARTHEW, P. and SLINGER, R. P. (1981) Diagnosis of sialodacryoadenitis virus infection of rats in a virulent enzootic outbreak, *Laboratory Animals*, **15**, 339–42.

CHIU, T. and CHEN, H. C. (1986) Spontaneous basophilic hypertrophic foci of the parotid glands in rats and mice, *Veterinary Pathology*, **23**, 606–9.

COLEMAN, G. L., BARTHOLD, S. W., OSBALDISTON, G. W., FOSTER, S. J. and JONAS, A. M. (1977) Pathological changes during aging in barrier-reared Fischer 344 male rats, *Journal of Gerontology*, **32**, 258–78.

CRAIN, R. C. (1958) Spontaneous tumours in the Rochester strain of Wistar rat, *American Journal of Pathology*, **34**, 311–35.

DAVIDSOHN, I., TAKAHASHI, T. and LEE, C. L. (1963) Liver infarction in mice following injection of antierythrocyte serum, *Federated Proceedings*, **22**, 545–8.

DAWE, C. J. (1979) Tumours of the salivary and lacrymal glands, nasal fossa and maxillary sinuses, in TURUSOV, V. S. (Ed.), *Pathology of Tumours in Laboratory Animals*, Vol. 2, *Tumours of the Mouse*, pp. 91–133, Lyon: IARC.

DEAN, D. H. and HIRAMOTO, R. N. (1984) Decreased plasma testosterone in desalivated male rats, *Canadian Journal of Physiology and Pharmacology*, **62**, 565–8.

DIDIER, R., REMESY, C., DEMIGNE, C. and FAFOURNOUX, P. (1985) Hepatic proliferation of mitochondria in response to a high protein diet, *Nutrition Research*, **5**, 1093–102.

ELWELL, M. R. and LEININGER, J. R. (1990) Tumours of the salivary and lacrymal glands, rat, in TURUSOV, V. and MOHR, U. (Eds), *Pathology of Tumours of Laboratory Animals*, Vol. 1, *Tumours of the Rat*, 2nd edition, pp. 89–107, Lyon: IARC.

ERNST, H. and MOHR, U. (1991) Ameloblastic odontoma of the mandible, rat, in JONES, T. C., MOHR, U. and HUNT, R. D. (Eds), *Pathology of Laboratory Animals: Cardiovascular and Musculoskeletal System*, pp. 218–24, Berlin: Springer-Verlag.

FITZGERALD, J. E. (1987) Ameloblastic odontoma in the Wistar rat, *Toxicologic Pathology*, **15**, 479–81.

FUKUSHIMA, S. and ITO, N. (1985) Papilloma and squamous cell carcinoma, forestomach, rat, in JONES, T. C., MOHR, U. and HUNT, R. D. (Eds), *Digestive System*, pp. 289–95, Berlin and Heidelberg: Springer-Verlag.

GOODMAN, D. G., WARD, J. M. and SQUIRE, R. A. (1979) Neoplastic and non-neoplastic lesions in aging F344 rats, *Toxicology and Applied Pharmacology*, **48**, 433–77.

GOODMAN, D. G., WARD, J. M., SQUIRE, R. A., CHU, K. C. and LINHART, M. S. (1980) Neoplastic and non-neoplastic lesions in aging Osborne-Mendel rats, *Toxicology and Applied Pharmacology*, **55**, 433–47.

GREAVES, P. and FACCINI, J. M. (1984) Digestive system, in *Rat Histopathology*, A glossary for use in toxicity and carcinogenicity studies, pp. 86–143, Amsterdam: Elsevier.

HARKNESS, J. E. and FERGUSON, F. G. (1979) Idiopathic megaoesophagus in rat, *Laboratory Animal Science*, **29**, 495–8.

HAYNES, D., HALL, P. and CLARK, D. (1983) A glycogen storage disease in rats. Morphological and biochemical investigations, *Virchows Archiv* [B], **42**, 289–301.

IMAI, K. and YOSHIMURA, S. (1988) Spontaneous tumours in Sprague-Dawley (CD:Crj) rats, *Journal of Toxicologic Pathology*, **1**, 7–12.

IRISARRI, E. and HOLLANDER, C. F. (1994) Aging of the liver, in MOHR, U., DUNGWORTH, D. L. and CAPEN, C. C. (Eds), *Pathobiology of the Aging Rat*, Vol. 2, pp. 341–9, Washington: ILSI Press.

IWATA, H., HIROUCHI, Y., KOIKE, Y. and YAMAKAWA, S. (1991) Historical control data on non-neoplastic and neoplastic lesions in F344/DuCrj rats, *Journal of Toxicologic Pathology*, **4**, 1–24.

JACOBY, R. Q., BHATT, P. N. and JONAS, A. M. (1979) Viral diseases, in BAKER, H. J., LINDSEY, J. R. and WEISBROTH, S. H. (Eds), *The Laboratory Rat*, Vol. 1, *Biology and Diseases*, pp. 271–306, New York: Academic Press.

KENDRY, G. and ROE, F. J. C. (1969) Histopathological changes in the pancreas of laboratory rats, *Laboratory Animals*, **3**, 207–20.

KLEIN-SZANTO, A. J. P., MARTIN, D. and SEGA, M. (1982) Hyperkeratinization and hyperplasia of the forestomach epithelium in vitamin A deficient rats, *Virchows Archiv* [B], **40**, 387–94.

KNOOK, D. L. (1982) Organ ageing in relation to cellular ageing, in VIIDIK, A. A. (Ed.), *Lectures on Gerontology*, pp. 213–61, London: Academic Press.

KOCIBA, R. J. and KEYES, D. G. (1985) Squamous cell carcinoma, tongue, rat, in JONES, T. C., MOHR, U. and HUNT, R. D. (Eds), *Digestive System*, Monograph on Pathology of Laboratory Animals, pp. 255–9, Berlin: Springer-Verlag.

KOUNTOURAS, J., BILLING, B. H. and SCHEUER, H. J. (1984) Prolonged bile duct obstruction: a new experimental model for cirrhosis in the rat, *British Journal of Experimental Pathology*, **65**, 305–11.

LEE, K. P. (1983) Peliosis hepatis-like lesions in aging rats, *Veterinary Pathology*, **20**, 410–23.

LEWIS, D. J., CHERRY, C. P. and GIBSON, W. A. (1980) Ameloblastoma (adamantinoma) of the mandible in the rat, *Journal of Comparative Pathology*, **90**, 379–84.

LIU, F. T. Y., LIN, H. S. and ZULLO, T. G. (1969) Effect of graded doses of estradiol benzoate on dental caries and salivary glands in female rats, *Journal of Dental Research*, **48**(3), 485.

MAEKAWA, A. (1994) Changes in the intestine, in MOHR, U., DUNGWORTH, D. L. and CAPEN, C. C. (Eds), *Pathobiology of the Aging Rat*, pp. 333–9, Washington: ILSI Press.

MAEKAWA, A., KUROKAWA, Y., TAKAHASHI, M., KOKUBO, T., OGIU, T., ONODERA, H., TANIGAWA, H., OHNO, Y., FURUKAWA, F. and HAYASHI, Y. (1983a) Spontaneous tumours in F344/DuCrj rats, *Gann*, **74**, 365–72.

MAEKAWA, A., KUROKAWA, Y., TAKAHASHI, M. and KOKUBO, T. (1983b) Neoplastic and non-neoplsatic diseases in aging Slc:Wistar rats, *The Journal of Toxicological Sciences*, **8**, 279–90.

MAITA, K., HIRANO, M., HARADA, T., MITSUMORI, K. (1986) An outbreak of esophagectasis in F344 rats, *Japanese Journal of Veterinary Science*, **48**, 539–46.

MARONPOT, R. R. (1990) Tumours of the liver, in STINSON, S. F., SCHULLER, H. M., REZNIK, G. (Eds), *Atlas of Tumour Pathology of the Fischer Rat*, pp. 193–220, Boca Raton, Florida: CRC Press.

NAGAYO, T. (1973) Tumours of the stomach, in TURUSOV, V. S. (Ed.), *Pathology of Tumours in Laboratory Rats*, pp. 88–101, Lyon: IARC.

POZHARISSKI, K. M. (1973a) Tumours of the oesophagus, in TURUSOV V.S. (Ed.), *Pathology of Tumours in Laboratory Rats*, Vol. 1, *Tumours of the Rat*, pp. 87–100, Lyon: IARC.

POZHARISSKI, K. M. (1973b) Tumours of the intestines, in TURUSOV, V. S. (Ed.), *Pathology of Tumours in Laboratory Rats*, Vol. 1, *Tumours of the Rat*, pp. 119–40, Lyon: IARC.

ROBERTS, J. C., MCCROSSAN, M. V. and JONES, H. B. (1990) The case for perfusion fixation of large tissue samples for ultrastructural pathology, *Ultrastructural Pathology*, **14**, 177–91.

ROBINSON, M. (1985) Dietary related periodontitis and oro-nasal fistulation in rats, *Journal of Comparative Pathology*, **95**, 489–98.

ROE, F. J. C. and ROBERTS, J. B. D. (1973) Tumours of the pancreas, in TURUSOV, V. S. (Ed.), *Pathology of Tumours in Laboratory Animals*, Vol. 1 *Tumours of the Rat*, pp. 141–50, Lyon: IARC.

ROWLATT, U. F. (1967) Neoplasms of the alimentary canal of rats and mice, in COTCHIN, E. and ROE, F. J. C. (Eds), *Pathology of Laboratory Rats and Mice*, pp. 57–84, Oxford: Blackwell Scientific.

RUBEN, Z., ROHRBACHER, E. and MILLER, J. E. (1983) Esophageal impaction in BHE rats, *Laboratory Animal Science*, **33**, 63–5.

SCHAUER, A. and KUNZE, E. (1976) Tumours of the liver, in TURUSOV, V. S. (Ed.), *Pathology of Tumours in Laboratory Animals*, Vol. 1, *Tumours of the Rat*, pp. 41–72, Lyon: IARC.

SCHMUCKER, D. L. (1990) Hepatocyte ultrastructure and aging, *Journal of Electron Microscopy Technique*, **14**, 106–25.

STEWART, H. L., WILLIAMS, G., KEYSSER, C. H., LOMBARD, L. S. and MONTALI, R. J. (1980) Histologic typing of liver tumours of the rat, *Journal of the National Cancer Institute*, **64**, 179–205.

STINSON, S. F. (1990) Spontaneous tumours in Fischer rats, in STINSON, S. F., SCHULLER, H. M. and REZNIK, G. (Eds), *Atlas of Pathology of the Fischer Rat*, pp. 2–17, Boca Raton, Florida: CRC Press.

STINSON, S. F. and KOVATCH, R. M. (1990) Tumours of the upper digestive tract (oral cavity, esophagus, forestomach), in STINSON, S. F., SCHULLER, H. M. and REZNIK, G. (Eds), *Atlas of Tumor Pathology of the Fischer Rat*, pp. 70–93, Boca Raton, Florida: CRC Press.

TAKAHASHI, M. and HASEGAWA, R. (1990) Tumours of the stomach, in JONES, V. and MOHR, U. (Eds), *Pathology of Tumours in Laboratory Animals*, Vol. 1, *Tumours of the Rat*, 2nd Edition, pp. 129–57, Lyon: IARC.

TATEMATSU, M. and IMAIDA, K. (1990a) Tumours of the glandular stomach, in STINSON, S. F., SCHULLER, H. M. and REZNIK, G. (Eds). *Atlas of Tumour Pathology of the Fischer Rat*, pp. 95–116, Boca Raton, Florida: CRC Press.

TATEMATSU, M. and IMAIDA, K. (1990b) Tumours of the small intestine, in STINSON, S. F., SCHULLER, H. M. and REZNIK, G. (Eds), *Atlas of Tumour Pathology of the Fischer Rat*, pp. 117–32, Boca Raton, Florida: CRC Press.

TUCKER, M. J. (1975) Carcinogenic action of quinoxaline 1,4 dioxide in rats, *Journal of the National Cancer Institute*, **55**, 137–45.

TUCKER, M. J. and ORTON, T. C. (1995) *Comparative Toxicology of Hypolipidaemic Fibrates*, Methylclofenapate, pp. 23–58, London: Taylor & Francis.

WAKE, K. (1974) Development of vitamin A rich lipid droplets in multivesicular bodies of rat liver stellate cells, *The Journal of Cell Biology*, **63**, 683–91.

WARD, J. M. (1981) Morphology of foci of altered hepatocytes and naturally occurring hepatocellular tumours in F344 rats, *Virchows Archiv [A]*, **390**, 339–45.

YU, B., MASORO, E. J., MURATA, I., BERTRAND, H. A. and LYND, F. T. (1982) Life span study of SPF Fischer 344 male rats fed ad libitum or restricted diets: longevity, growth, lean body mass and disease, *Journal of Gerontology*, **37**, 130–41.

ZIMMERMAN, H. J. (1978) Classification of hepatotoxins and mechanisms of toxicity, in *Hepatotoxicity. The Adverse Effects of Drugs and Other Chemicals on the Liver*, pp. 91–121, New York: Appleton Century Crofts.

5

The Urinary System

5.1 Kidney

The kidneys of the rat are still developing at birth, and the number of glomeruli increases until 100 days. The final number of nephrons is strain dependant and varies between 20 000 and 40 000 (Arataki, 1926). The rat kidney is unilobular (i.e. it has a single pyramid), unlike the human kidney which has 10–14 lobules. The cortex forms a cap over the medulla and the tip (papilla) projects into the renal sinus.

5.1.1 Non-neoplastic Changes

Congenital anomalies

Both kidneys are sliced at necropsy, prior to fixation, in transverse section above and below the hilus so that the histological section passes through cortex, medulla and papilla. Congenital abnormalities of the kidney are rare in the AP rat (<1 per cent) and include polycystic kidney, hypoplastic kidney, agenesis, and absent renal blood vessels. Polycystic kidney is usually fatal within a few months, but the other abnormalities have been incidental findings during histological examination.

Kidney weights

The weight of the kidneys in AP rats is shown in Table 5.1. The absolute weight of the kidney increases up to 34 weeks but shows little change after this time. The relative kidney weights (kidney weights as a percentage of body

Table 5.1 Kidney weights in the AP rat

Age (weeks)	Number/ sex	Mean kidney weights			
		Weight (g)		Relative weight (% body weight)	
		Males	Females	Males	Females
12	10	2.65	1.70	0.73	0.76
34	20	3.71	2.50	0.61	0.71
58	25	3.79	2.51	0.52	0.64

weight) are always greater in females, but show a decline in both sexes from 12 weeks reflecting the increases in non-lean body weight. This pattern of growth is different from that described for SD rats by Owen and Heywood (1986) who reported a continuous increase in kidney weight, so that the relative weight remains unchanged up to 108 weeks of life. Tuma *et al.* (1985) reported that female F344 rat kidneys increase in weight by 23 to 30 per cent between 12 and 24 months although lean tissue remains unchanged. They concluded that this indicated renal hypertrophy.

Chronic progressive glomerulonephropathy

The most important spontaneous disease of the kidney is chronic progressive glomerulonephropathy (CPGN) which is also known as chronic progressive nephrosis, progressive glomerulosclerosis, and old rat nephropathy. After neoplastic disease it is probably the most common cause of morbidity and mortality in most strains of rats although the age of onset and the severity varies with strain (Bolton *et al.*, 1976; Coleman *et al.*, 1977; Gray *et al.*, 1982; Solleveld and Boorman, 1986). In the AP rat, as in most strains, it is more common in males (Gray, 1977). The onset in the AP rat is quite early with males of 6 months showing minimal changes consisting of scattered sclerotic glomeruli, a few foci of tubules dilated with eosinophilic (hyaline) casts and small mononuclear cell infiltrates in the interstitial tissue (Figure 38). These changes are sufficient to cause some functional disturbance, chiefly an increased permeability of the glomerular basement membrane, and a proteinurea which increases with time. The major component of this proteinurea in young males is α_{2u}-globulin; in older animals of both sexes it is albumin (Bolton *et al.*, 1976). At 12 months the majority of males show minimal to mild changes. From 18 months CPGN is a significant factor in the mortality of the AP rat, although when a neoplastic disease, such as a large pituitary tumour, is also present it is not possible to determine if one disease is ultimately responsible for the moribund state of the animal. At all times CPGN is less severe in females and only occasional animals will show the most severe form. In the final stages of the

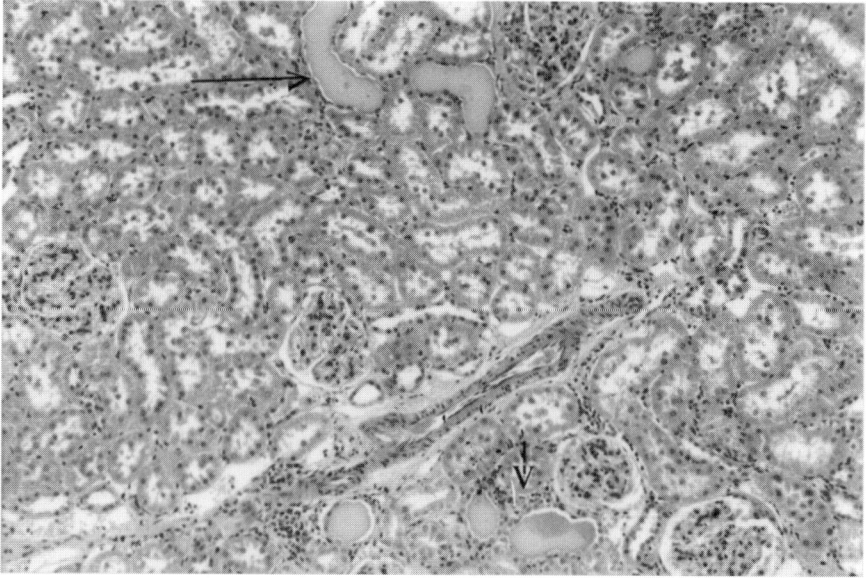

Figure 38 Early stage of chronic progressive glomerulonephropathy (CPGN) in a male AP rat aged 8 months showing a few dilated tubules (←) with casts. ×80, H&E

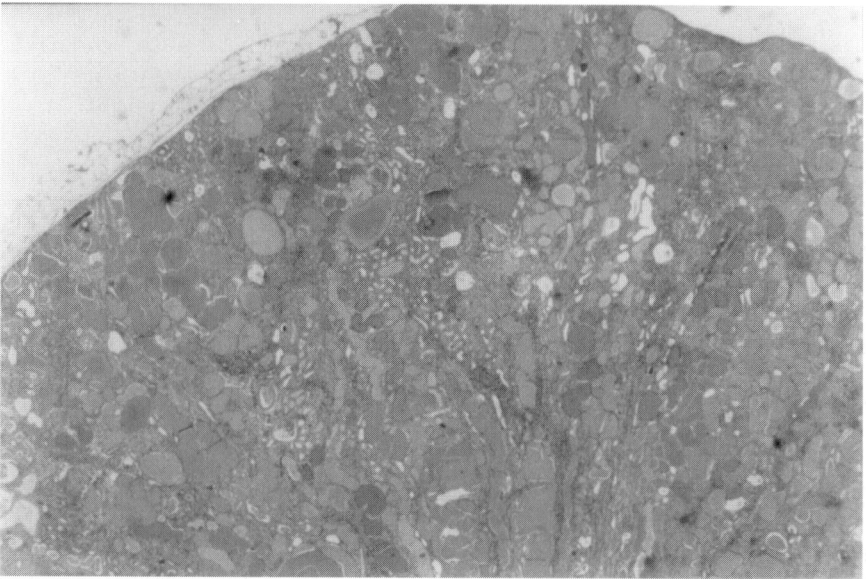

Figure 39 Severe CPGN in a male AP rat aged 20 months. ×8, H&E

83

disease there are several non-specific clinical signs including loss of body weight and body condition with urine staining of urinogenital fur, and often red porphyrin staining around the eyes. The animals are lethargic and disinclined to eat. At necropsy the kidneys are enlarged and yellow in colour sometimes with an 'orange peel' surface; cystic tubules may be large enough to be visible macroscopically. No normal tissue remains within the kidney (Figure 39) which shows a range of changes including varying degrees of tubular dilatation and cast formation with severe, universal, glomerular sclerosis or dilatation of Bowman's space with atrophic glomerular tuft remnants (Figure 40). Tubules may be atrophied with gross thickening of basement membranes and interstitial inflammation or fibrosis (Figure 41); nuclear crowding may be present in some proximal tubules (Figure 42) and there may be calcification of the basement membrane of tubules (Figure 43). Calcification is only present if the rat has developed secondary hyperparathyroidism. This condition affects up to 5 per cent of males at 2 years. They show marked hyperplasia of the parathyroid glands and widespread metastatic calcification of many organs, including the interstitial tissue of the lung, the walls of coronary vessels and myocardium in the heart, and in the muscle and mucosa of the gastro-intestinal tract. The bones may show changes similar to osteitis fibrosa cystica with fibrosis of marrow and resorption of the bone by endosteal osteoclasts; microscopic cysts in bone or marrow are often, but not always, present.

The pathogenesis of CPGN has not been elucidated completely but there is a clear relationship to proteinurea. Rats which develop a severe proteinurea also

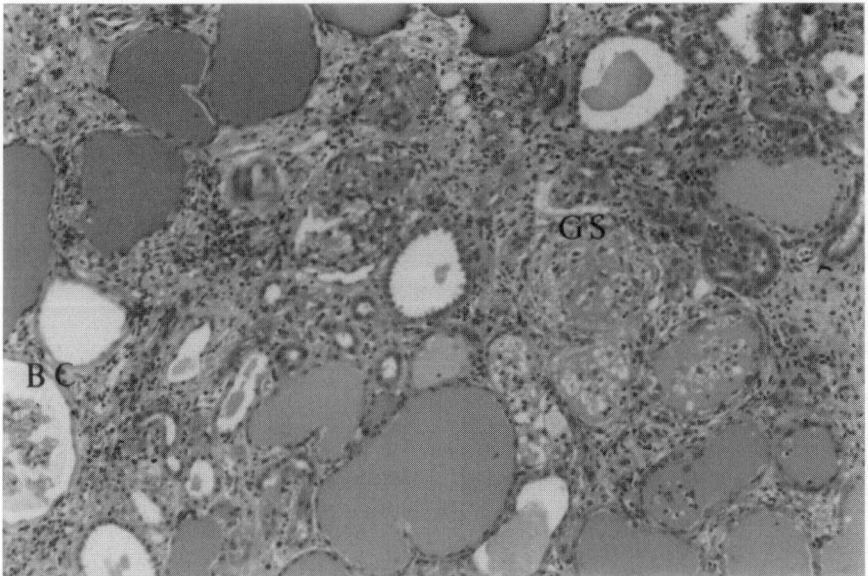

Figure 40 Severe CPGN: glomerular sclerosis (GS) with dilatation of Bowman's capsule (BC) and dilated tubules with proteinaceous casts. ×80, H&E

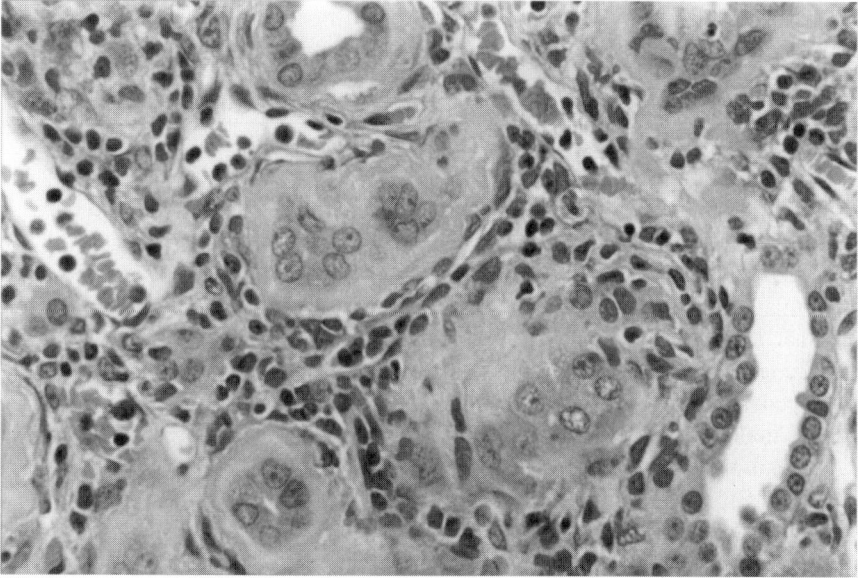

Figure 41 Severe CPGN: atrophic cortical tubules with thickened basement membranes and interstitial inflammation. ×128, H&E

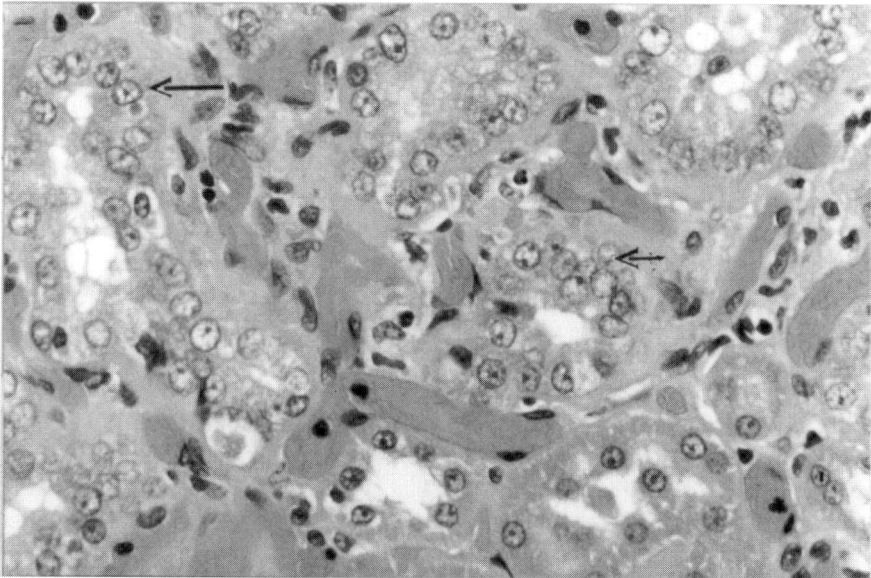

Figure 42 Severe CPGN: proximal tubules showing nuclear crowding. ×128, H&E

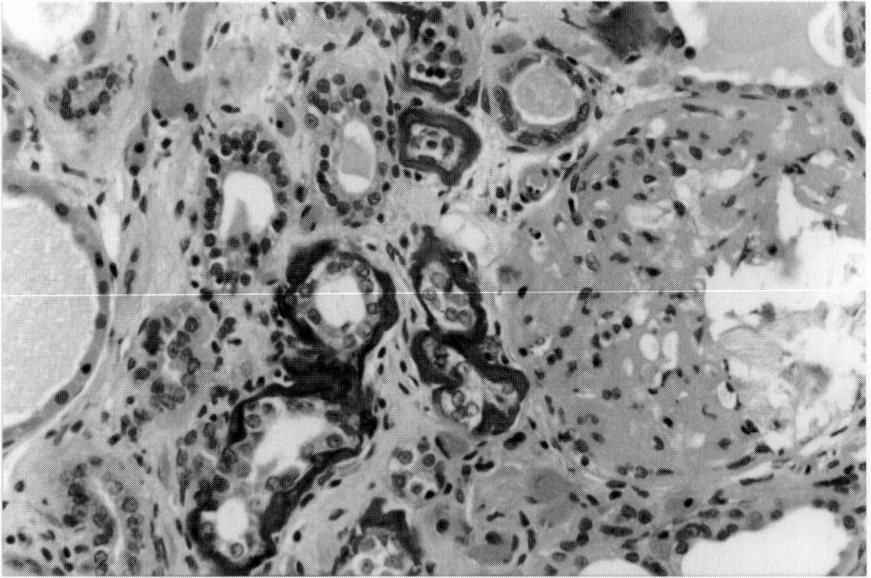

Figure 43   Severe CPGN: tubules with calcified basement membranes. ×128, H&E

develop CPGN, but rats with only a low level proteinurea do not (Owen and Heywood, 1986). Laboratory diets which have a high protein level are associated with development of CPGN, while low protein diets or food restriction reduce the incidence (Saxton and Kimball, 1941; Tucker *et al.*, 1976; Everitt *et al.*, 1982; Dodane *et al.*, 1991). It has been suggested by Tapp *et al.* (1989) that it is reduced caloric intake not the reduced protein level which affects the development of CPGN. In their rat ablation model, restriction of caloric intake reduced the incidence of CPGN, while restriction of protein level, without a reduction of calorific value, had no effect on the incidence of the disease. Reduced food intake (by 20 per cent) in the AP rat has resulted in a marked reduction in the incidence of CPGN (unpublished observations). The age-related proteinurea varies with the strain of rat and correlates with the incidence of CPGN. F344 rats show a lower level of proteinurea than SD rats (Short and Goldstein, 1994). In all strains, levels of proteinurea are higher in males.

Other factors have also been implicated, e.g. administration of exogenous prolactin precipitates CPGN (Richardson and Luginbüehl, 1976), while administration of dopamine agonists such as bromocriptine inhibit the disease (Richardson *et al.*, 1984). It is known that, in the rat, serum prolactin levels increase with age (Stefaneanu and Kovacs, 1994), due to the reduction in secretion of the inhibitory hormone dopamine. Dietary restriction reduces the age-related increase of serum prolactin in female rats (Merry *et al.*, 1985; Atterwill *et al.*, 1989) but not in males, yet dietary restriction reduces the incidence of CPGN in males. In SD rats, levels of growth hormone, not prolactin, have been shown to correlate with the severity of nephropathy (Goya *et al.*, 1991). There is, clearly, a complex inter-relationship between hormone levels, food intake and CPGN, which indicates that there is unlikely to be a single causal factor in the development of the disease.

Protein droplets

Protein (hyaline) droplets in the S2 segment of the proximal tubule are universal in the young male AP rat, but very uncommon in the female. They have been examined by electron microscopy by Maunsbach *et al.* (1962) and shown to be crystalloid phagolysosomal structures. They are found in the kidneys of most strains of rat but are more prominent in the Wistar and F344 than in SD rats (Short and Goldstein, 1994). The chief component of the droplets is α_{2u}-globulin which is a low molecular weight protein synthesised by the liver at puberty (Kanerva *et al.*, 1987). The NCI-Black-Reiter male rat does not show these protein droplets in the kidney because the liver does not synthesise α_{2u}-globulin (Ridder *et al.*, 1988). Liver synthesis and renal droplets decrease with age, with no detectable amounts in rats of 26 months of age (Murty *et al.*, 1988). In CPGN there may be protein droplets in the tubules but these are thought to be albumin that has leaked through the damaged glomeruli.

Hydronephrosis

Slight dilatation of the renal pelvis, without any evidence of renal parenchymal damage, is a quite common finding, usually on the right side. This has been attributed to compression of the ureter by the overlying spermatic or ovarian artery (Burton *et al.*, 1979). In severe pelvic dilatation (hydronephrosis) the kidney tissue may be reduced, by pressure, to a thin rim around the dilated pelvis, and there is tubular degeneration with interstitial inflammation and fibrosis indicative of obstruction. If the condition is unilateral, which is usually the case, the contralateral kidney may enlarge to compensate for the loss of functional renal tissue. In the AP rat the highest incidence of mild pelvic dilatation which has been observed is 53 per cent, but severe hydronephrosis has only been found in 1 to 2 per cent. A high incidence of the severe lesion was reported in BN/Bi rats (Burek, 1978) where the incidence was 43 and 39 per cent in males and females, respectively. In this strain it was associated with tumours or urolithiasis.

Inflammation

Pyelonephritis is a rare condition in the AP rat, with an overall incidence less than 1 per cent. Affected kidneys may show ulceration and necrosis of the papilla tip and radial foci of tubules containing a suppurative exudate. The usual cause is an ascending infection from the bladder. Minor inflammatory conditions of the pelvis, unrelated to CPGN or hydronephrosis are quite common but, as they do not progress to more severe conditions, they are not considered important. This includes foci of inflammatory cells under the urothelium lining the pelvis, and small exudates into the pelvic cavity. They may be accompanied by focal or minimal diffuse hyperplasia of the epithelium lining the pelvis. A review of pyelonephritis has been published by Duprat and Burek (1986). It is rare in F344 and SD rats (Coleman *et al.*, 1977; Montgomery and Seely, 1990), probably a reflection of the SPF status of most present-day rat colonies.

Nephrocalcinosis

Nephrocalcinosis includes pelvic urolithiasis, which is rare (<1 per cent) in AP rats, and intratubular microlithiasis (Figure 44), which is common in females (with a maximum 100 per cent incidence) and uncommon in males (where the maximum incidence observed is 5 per cent). This microlithiasis involves small deposits of calcium phosphate (microliths) in the S3 segment of the proximal tubule at the outer cortico-medullary junction. The aetiology of this microlithiasis is considered to be related to several factors – including oestrogen levels (Armstrong and Horsley, 1986) and variations in dietary levels of minerals, including the ratio of calcium and phosphorus (Clapp *et al.*, 1982; Ritskes-Hoitinga *et al.*, 1989). The incidence of intratubular microlithiasis in other strains of rat includes a 100 per cent incidence in female F344 rats at 12 months, but a lower and later incidence of 54 per cent in female SD rats at 24 months. The incidence in male rats varies from less than 10 per cent in Wistar

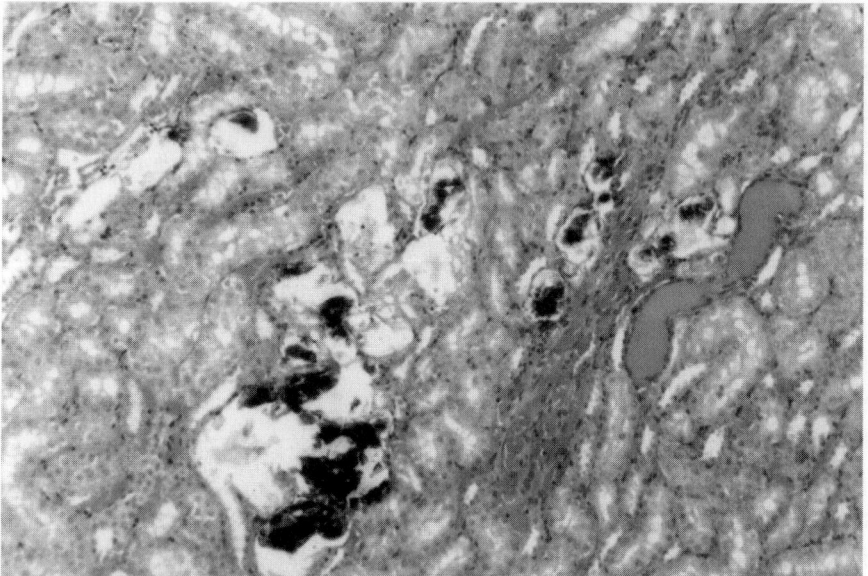

Figure 44   Intratubular microlithiasis in a 4 month old female AP rat. ×80, H&E

and SD strains to 50 per cent in F344 rats (Short and Goldstein, 1994; Peter *et al.*, 1986).

Pigmentation

Another kidney condition is minimal to mild pigment deposits in proximal tubules, which is found in most rats over 18 months of age. Special staining techniques have shown that the pigment is usually lipofuschin, but occasionally ceroid or haemosiderin. Similar deposits have been observed in F344 rats (Coleman *et al.*, 1977).

Oncocytes

Hypertrophied renal tubular epithelial cells with an intense eosinophilic cytoplasm have been termed oncocytes by some workers. In the AP rat they are rare and are usually seen in single tubules, occasionally in a focus of several tubules, and they are most frequent in kidneys showing severe CPGN. Tsuda and Krieg (1994) consider these oncocytic foci to be adenomas but others have suggested that they represent a disease of mitochondria (Tremblay, 1969). Some SD strains have a 100 per cent incidence of oncocytic cells (Peter *et al.*, 1986); this may indicate that oncocytes are not a pre-neoplastic change and the significance is unknown and their cytogenesis unresolved (Yamada *et al.*, 1988).

Infarcts

Infarcts are also rare (less than 1 per cent); they are visible macroscopically as capsular depressions, and microscopically appear as a wedge-shaped area of tubular atrophy and fibrosis. They are reported to be uncommon in all strains in the review of infarcts by Montgomery (1986).

5.1.2 Neoplastic Changes

Tumours of the kidney are rare in the AP rat and have not been observed in the majority of carcinogenicity studies. The highest incidence in any study is 1.5 per cent (9/600 control animals). There is no difference in incidence between the sexes and, with the exception of the nephroblastomas, all tumours occurred in animals more than 20 months of age. In spite of their rarity there is an extensive range of histological types, and these are shown Table 5.2.

Tubular epithelial tumours

The most common tubular epithelial tumour is the tubular adenoma. These tumours were all incidental findings during histological examination in animals

Table 5.2 Histological types of kidney tumours found in the AP rat

Histological type	Number observed[a]	
	Males	Females
Tubular adenoma	3	3
Tubular adenocarcinoma	0	3
Tubular solid carcinoma	0	1
Transitional cell papilloma	2	0
Transitional cell carcinoma	1	2
Lipoma	0	3
Liposarcoma	1	2

[a] Tumour incidence from a database of 8880 AP rats (including 2800 males and 2500 females in 2 year studies) used in toxicology studies between 1960 and 1994.

killed at scheduled termination of 2 year studies. They were small, well-differentiated tumours showing clear tubular patterns of cells with an eosinophilic cytoplasm. Tubular adenocarcinomas were larger and less well-differentiated tumours, but with recognisable tubular patterns and some local invasion, but none was the cause of death in the animal. One undifferentiated tumour was classified as a solid (clear cell) adenocarcinoma. It was a large tumour occupying the whole of one kidney and histologically showed no tubular pattern but sheets of epithelial cells with a vacuolated cytoplasm. Bannasch and Zerban (1986) consider that the malignant tumours are a progression from the adenomas.

The transitional cell papilloma arose in the renal pelvis and was a simple tumour with a connective tissue core covered by a transitional epithelium several cells thick. The carcinomas were poorly defined and highly invasive tumours. They consisted of small islands of transitional cells in a connective tissue stroma, or solid masses of pleomorphic cells. All showed areas of squamous metaplasia, while some showed such extensive squamous elements that a diagnosis of squamous cell carcinoma seemed more appropriate.

Mesenchymal and embryonal tumours

The three nephroblastomas were all found in animals which were killed because of weight loss and poor body condition before they reached 6 months of age. Macroscopically the involved kidney was enlarged and adherent to adjacent tissues. The histological appearance is similar to that described by other workers (Cardesa and Ribalta, 1986; Hard, 1990), with nests of embryonic cells often forming primitive glomeruli and tubules. The nests are surrounded by an extensive connective tissue stroma. In general the tumours showed a high mitotic rate, and their appearance is so characteristic that they do not present any difficulties in diagnosis. This is a rare tumour in all strains except the Upjohn SD rat (Mesfin and Breech, 1996). Lipomas and liposarcomas had the characteristic morphology of these tumours elsewhere in the body. Secondary (metastatic) tumours seen in the kidney include leukaemias and other lymphomas, and an osteosarcoma. A low incidence of kidney tumours is found in all other strains such as the F344 (Reznik, 1990), Osborne-Mendel (Goodman *et al.*, 1980) and Wistar (Maita *et al.*, 1979). Kidney tumours (tubular carcinomas) have been induced in the AP rat by quinoxaline 1,4 dioxide (Tucker, 1975) and benign tubular adenomas by streptozotocin (unpublished data).

5.2 Ureter

5.2.1 *Non-neoplastic Changes*

The ureter has not been examined routinely in the AP rat, so only macroscopically abnormal ureters have been examined and these have been few. The only abnormalities observed have been obstructive dilatation, epithelial hyperplasia and mild ureteritis.

5.2.2 *Neoplastic Changes*

No tumours have been found in the ureter in the AP rat. This also applies to most laboratory rat strains except for the Brown Norway rat, which had a high incidence of 20 per cent in females and 6 per cent in males; 6 per cent of these tumours had lung metastases (Boorman and Hollander, 1974; Boorman *et al.*, 1977).

5.3 Urinary Bladder

5.3.1 *Non-neoplastic Changes*

For many years the urinary bladder of the AP rat was examined after *in situ* fixation. A small amount of fixative was introduced into the bladder via the urethra and left for 10 minutes. After this time the neck of the bladder was tied. The bladder was removed and examined by transmitted light and then the wall opened and the inner surface examined before the organ was put into fixative. This technique was used because we had identified several bladder carcinogens in carcinogenicity studies. After many years no abnormality had been detected using this method and less than 1 per cent of animals showed any significant histological changes in the bladder. A decision was made to remove the bladder in a similar manner to other organs, and fixation thereafter was by immersion.

Inflammation

Spontaneous changes are also uncommon in the urinary bladder. Inflammatory cell infiltrates, usually mononuclear cell, may be scattered in the wall of the bladder but frank cystitis is not a frequent finding. It has been seen in less than ten animals in the database and in all of these animals it was secondary to local irritation caused by large calculi (stones).

In addition to the cystitis, diffuse hyperplasia of the lining epithelium was also present (Figure 45). This low incidence of calculi and inflammatory lesions appears to be true of other Wistar rat colonies and also of other rat strains (Kihlstrum and Clements, 1969; Burek, 1978; Goodman *et al.*, 1979).

Coagula

The most common finding is the presence of proteinaceous plugs (coagula) which are present in male rats of all ages and are often sufficiently large to fill the lumen of the bladder. The coagulum is composed of seminal fluid and exfoliated urothelial cells, and the incidence in male AP rats varies between 0 and 20 per cent. This is in accord with the incidence reported for other strains (Chowaniec and Hicks, 1979; Lee, 1986). The formation of coagula is thought to be the result of abnormal ejaculation.

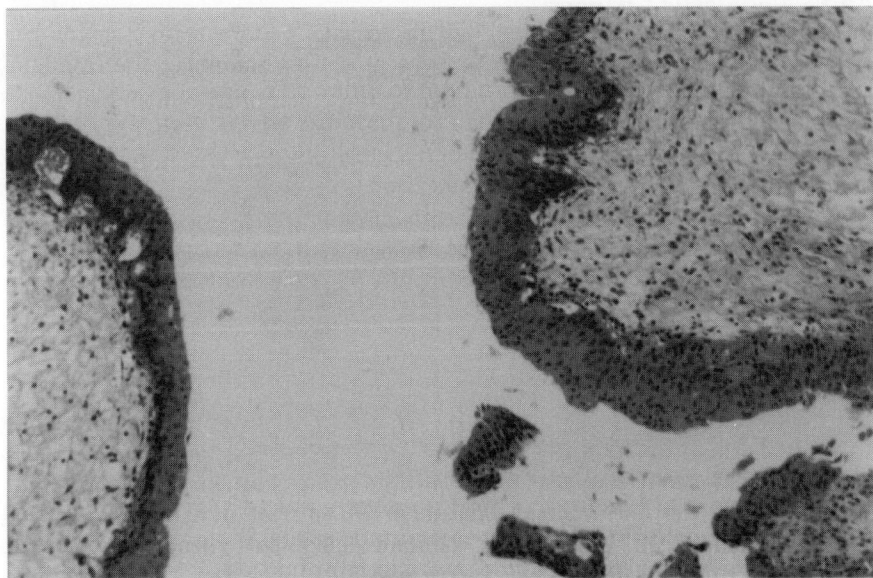

Figure 45   Urinary bladder: diffuse hyperplasia of the transitional cell epithelium and sub-epithelial inflammation. ×80, H&E

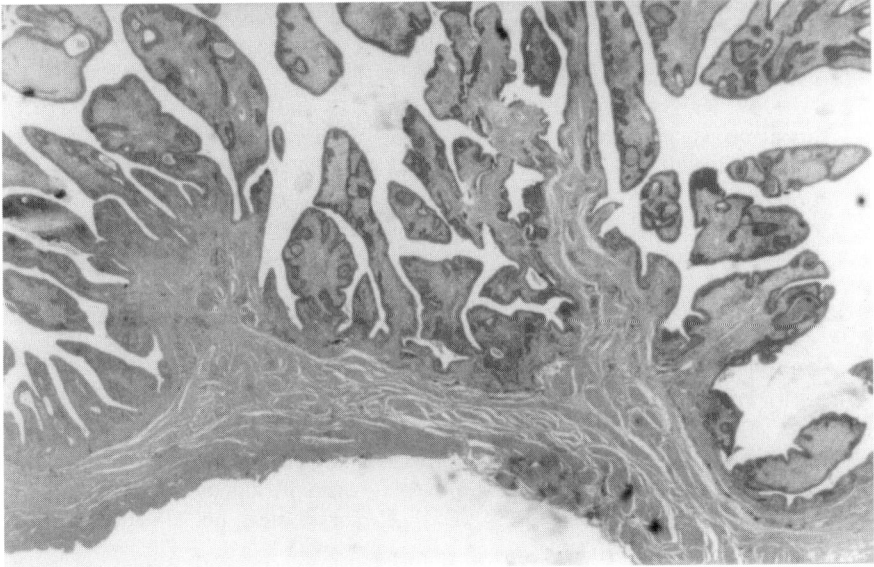

Figure 46 Sessile transitional cell papilloma of the urinary bladder in a female AP rat aged 26 months. ×8, H&E

5.3.2 *Neoplastic Diseases*

Spontaneous bladder tumours are very rare in the AP rat and include three transitional cell papillomas (Figure 46): two in males aged 26 and 29 months (one with a large bladder stone) and one in a female of 26 months. There were two transitional cell carcinomas in females aged 10 and 26 months, and a leiomyoma in a female aged 26 months. Bladder tumours are uncommon in most strains but a high incidence was seen in BN/Bi rats by Burek (1978). The mean age of his rats was 27 months (males) and 33 months (females). As few rat toxicology studies are maintained for this period of time, it may account for the low incidence reported by other workers (Goodman *et al.*, 1979; Goodman *et al.*, 1980).

5.4 References

ARATAKI, M. (1926) On the post natal growth of the kidney with significant reference to the number and size of glomeruli, *American Journal of Anatomy*, **36**, 399–436.

ARMSTRONG, S. and HORSLEY, H. J. (1986) A sex determined renal calcification in rats, *Nature*, **21**, 980.

ATTERWILL, C. K., BROWN, C. G., CONYBEARE, G., HOLLAND, C. W. and JONES, C. A. (1989) Relation between dopaminergic control of pituitary lactotroph function and deceleration of age-related changes in serum prolactin in diet-restricted rats, *Food and Cosmetic Toxicology*, **27**, 97–103.

BANNASCH, P. and ZERBAN, H. (1986) Renal cell adenoma and carcinoma, rat, in JONES, T. C., MOHR, U. and HUNT, R. D. (Eds), *Urinary System*, pp. 112–39, Berlin and Heidelberg: Springer-Verlag.

BOLTON, W. K., BENTON, F. R., MACLAY, J. G. and STURGILL, B. C. (1976) Spontaneous glomerulosclerosis in aging Sprague-Dawley rats. 1: Lesions associated with mesangial IgM deposits, *American Journal of Pathology*, **85**, 277–302.

BOORMAN, G. A. and HOLLANDER, C. F. (1974) High incidence of spontaneous urinary bladder and ureter tumours in the Brown Norway rat, *Journal of the National Cancer Institute*, **52**, 1005–8.

BOORMAN, G. A., BUREK, J. D. and HOLLANDER, C. F. (1977) Animal model: spontaneous urothelial tumors in BN/BiRij rats, *American Journal of Pathology*, **88**, 251–4.

BUREK, J. D. (1978) *Pathology of Aging Rats*, Florida: CRC Press.

BURTON, D. S., MARONPOT, R. R. and HOWARD, F. L. (1979) Frequency of hydronephrosis in Wistar rats, *Laboratory Animal Science*, **29**, 642–4.

CARDESA, A. and RIBALTA, T. (1986) Nephroblastoma, kidney, rat, in JONES, T. C., MOHR, U. and HUNT, R. D. (Eds), *Urinary System*, pp. 71–80, Berlin and Heidelberg: Springer-Verlag.

CHOWANIEC, W. H. and HICKS, R. M. (1979) Response of the rat to saccharin with particular reference to the urinary bladder, *British Journal of Cancer*, **39**, 355–75.

CLAPP, M. J. L., WADE, J. D. and SAMUELS, D. M. (1982) Control of nephrocalcinosis by manipulating the calcium:phosporus ratio in commercial rodent diets, *Laboratory Animals*, **16**, 130–2.

COLEMAN, G. L., BARTHOLD, S. W., OSBALDISTON, G. W., FOSTER, S. J. and JONAS, A. M. (1977) Pathological changes during aging in barrier-reared Fischer F344 male rats, *Journal of Gerontology*, **32**, 258–78.

DODANE, V., CHEVALIER, J., BARIETY, J., PRATZ, J. and CORMAN, B. (1991) Longitudinal study of solute excretion and glomerular ultrastructure in an experimental model of aging rats free of kidney disease, *Laboratory Investigation*, **64**, 377–91.

DUPRAT, P. and BUREK, J. D. (1986) Suppurative nephritis, pyelonephritis, rat, in JONES, T. C., MOHR, U. and HUNT, R. D. (Eds), *Urinary System*, pp. 219–24, Berlin and Heidelberg: Springer-Verlag.

EVERITT, A. V., PORTER, B. D., PARSONS, C. E. and READ, N. G. (1982) Effects of caloric intake and dietary composition on the development of proteinurea, age associated renal disease and longevity in the male rat, *Gerontology*, **28**, 168–75.

GOODMAN, D. G., WARD, J. M., SQUIRE, R. A., CHU, K. C. and LINHART, M. S. (1979) Neoplastic and non-neoplastic lesions in aging F344 rats, *Toxicology and Applied Pharmacology*, **48**, 237–48.

GOODMAN, D. G., WARD, G. B., SQUIRE, R. A., PAXTON, M. B., REICHARDT, W. D., CHU, K. C. and LINHART, M. S. (1980) Neoplastic and non-neoplastic lesions in the aging Osborne-Mendel rats, *Toxicology and Applied Pharmacology*, **55**, 433–47.

GOYA, R. G., CASTELLETO, L. and SOSA, Y. E. (1991) Plasma levels of growth hormone correlate with the severity of pathologic changes in the renal structure of aging rats, *Laboratory Investigation*, **64**, 29–34.

GRAY, J. E. (1977) Chronic progressive nephrosis in the albino rat, *CRC Critical Reviews in Toxicology*, **5**, 115–44.

GRAY, J. E., VAN ZWIETEN, M. J. and HOLLANDER, C. F. (1982) Early light microscopic changes of chronic progressive glomerular nephrosis in several strains of aging laboratory rats, *Journal of Gerontology*, **37**, 142–50.

HARD, G. C. (1990) Tumours of the kidney, renal pelvis and ureter, in TURUSOV, V. S. and MOHR, U. (Eds), *Pathology of Tumours in Laboratory Animals*, Vol 1, *Tumours of the Rat*, 2nd edition, pp. 301–44, Lyon: IARC.

KANERVA, R. L., MCCRAKEN, M. S., ALDEN, C. L. and STONE, L. C. (1987) Morphogenesis of decalin-induced renal alterations in the male rat, *Food and Chemicals Toxicology*, **25**, 53–61.

KIHLSTRUM, J. M. and CLEMENTS, G. R. (1969) Spontaneous pathologic findings in Long-Evans rats, *Laboratory Animal Care*, **19**, 710–15.

LEE, K. P. (1986) Ultrastructure of proteinaceous bladder plugs in male rats, *Laboratory Animal Science*, **36**, 671–7.

MAITA, K., MATSUNUMA, N., MASUDA, H. and SUZUKI, Y. (1979) The age-related tumour incidence in Wistar-Imamichi rat, *Experimental Animals* (Tokyo), **28**, 555–60.

MAUNSBACH, A. B., MADDDEN, S. C. and LATTA, H. (1962) Light and electronmicroscopic changes in proximal tubules of rats after administration of glucose, mannitol, sucrose or dextran, *Laboratory Investigation*, **11**, 421–32.

MERRY, B. J., HOLEHAN, A. M. and PHILLIPS, J. G. (1985) Modification of reproductive decline and lifespan by dietary manipulation in Sprague-Dawley rats, in LOFTS, B. and HOLMES, W. N., *Current Trends in Comparative Endocrinology*, Hong Kong: Hong Kong University Press.

MESFIN, G. M. and BREECH, K. T. (1996) Heritable nephroblastoma (Wilm's tumor) in the Upjohn Sprague Dawley rat, *Laboratory Animal Science*, **46**, 321–6.

MONTGOMERY, C. A. (1986) Infarction, kidney, rat, mouse, in JONES, T. C., MOHR, U. and HUNT, R. D. (Eds), *Urinary System*, pp. 179–83, Berlin: Springer-Verlag.

MONTGOMERY, C. A. and SEELY, J. C. (1990) Kidney, in BOORMAN, G. A., ELWELL, M. R., EUSTIS, S. L. and MONTGOMERY, C. A. (Eds), *Pathology of the Fischer Rat*, pp. 127–53, Orlando: Academic Press.

MURTY, C. V. R., OLSON, M. J., GARG, B. D. and ROY, A. K. (1988) Hydrocarbon-induced hyaline droplet nephropathy in male rats during senescence, *Toxicology and Applied Pharmacology*, **96**, 380–92.

OWEN, R. and HEYWOOD, R. (1986) Age-related variation in renal structure and function in Sprague-Dawley rats, *Toxicologic Pathology*, **14**(2), 158–67.

PETER, C. P., BUREK, J. D. and VAN ZWIETEN, M. J. (1986) Spontaneous nephropathies in rats, *Toxicologic Pathology*, **14**, 91–100.

REZNIK, G. (1990) Neoplasms of the kidney, in STINSON, S. F., SCHULLER, H. M. and REZNIK, G. (Eds), *Atlas of Tumor Pathology of the Fischer Rat*, pp. 227–36, Boca Raton, Florida: CRC Press.

RICHARDSON, B. P. and LUGINBUEHL, H.-R. (1976) The role of prolactin in the development of chronic progressive nephropathy in the rat, *Virchow's Archiv [A]*, **370**, 13–15.

RICHARDSON, B. P., TURKALJ, I. and FLUCKIGER, E. (1984) Bromocriptine, in LAURENCE, D. R., MCLEAN, A. E. M. and WEATHERALL, M. (Eds), *Safety Testing of New Drugs*, pp. 3–64, London: Academic Press.

RIDDER, G. M., VON BARGEN, E. C., PARKER, R. D. and ALDEN, C. L. (1988) Spontaneous and induced accumulation of α_{2u}-globulin in the kidney cortex of rats and mice, *Toxicologist*, **8**, 352.

RITSKES-HOITINGA, J., LEMMENS, A. G. and BEYNEN, A. C. (1989) Nutrition and kidney calcification in rats, *Laboratory Animals*, **23**, 313–18.

SAXTON, J. and KIMBALL, G. (1941) Relation of nephrosis and other diseases of albino rats to age and to modification of diet, *Archives of Pathology*, **32**, 951–65.

SHORT, B. G. and GOLDSTEIN, R. S. (1994) Nonneoplastic lesions in the kidney, in MOHR, U., DUNGWORTH, C. G. and CAPEN, C. C. (Eds), *Pathobiology of the Aging Rat*, Vol. 1, pp. 211–25, Washington: ILSI Press.

SOLLEVELD, H. A. and BOORMAN, G. A. (1986) Spontaneous renal lesions in five rat strains, *Toxicologic Pathology*, **14**, 168–74.

STEFANEANU, L. and KOVACS, K. (1994) Changes in structure and function of the pituitary, in MOHR, U., DUNGWORTH, D. L. and CAPEN, C. C. (Eds) *Pathobiology of the Aging Rat*, Vol. 2, pp. 173–91, Washington: ILSI Press.

TAPP, D. C., WORTHAM, W. G., ADDISON, J. F., HAMMONDS, D. N., BARNES, J. L. and VENKATACHALAM, M. A. (1989) Food restriction retards body growth and prevents end-stage renal pathology in remnant kidney of rats regardless of protein intake, *Laboratory Investigation*, **60**(10), 1184–90.

TREMBLAY, G. (1969) The oncocytes, in BAJUSZ, E and JASMIN, G. (Eds), *Methods and Achievements in Experimental Pathology*, Vol. 4, pp. 121–40, Basel: Karger.

TSUDA, H. and KRIEG, K. (1994) Neoplastic lesions in the kidney, in MOHR, U., DUNGWORTH, D. L. and CAPEN, C. C. (Eds), *Pathobiology of the Aging Rat*, pp. 226–40, Washington: ILSI Press.

TUCKER, M. J. (1975) Carcinogenic action of quinoxaline-1,4-dioxide in rats, *Journal of the National Cancer Institute*, **55**, 137–45.

TUCKER, S., MASON, R. and BEAUCHENE, R. (1976) Influence of diet and feed restriction on kidney function of aging male rats, *Journal of Gerontology*, **31**(3), 264–70.

TUMA, F. T., IRION, G. L., VAITHARE, U. S. and HEINEL, L. A. (1985) Age-related changes in regional blood flow in the rat, *American Journal of Physiology*, **249**(18), 485–91.

YAMADA, S., ASA, S. L. and KOVACS, K. (1988) Oncocytomas and null cell adenomas of the human pituitary: morphometric and in vivo functional comparison, *Virchows Archiv [A]*, **413**, 333–9.

6

The Cardiovascular System

6.1 Heart

6.1.1 *Non-neoplastic Changes*

In toxicology studies in the 1960s the heart of AP rats was sampled for histological examination by a transverse section taken through the ventricles just below the atria, but since the 1970s the sample has been a longitudinal section which passes through all four chambers and the papillary muscle.

Cardiac weights

The weight of the heart varies with age as shown in Table 6.1. The absolute weight of the heart increases between 12 and 34 weeks of age but shows little change thereafter. In contrast the relative weight (heart weight as per cent of body weight) decreases with time, reflecting the increase in body weight with time. Tanase *et al.* (1982) compared strains of rat and showed that there are considerable differences in the size of the heart among the 23 strains they reviewed. Differences in the same strain, from different suppliers, was noted by Campbell and Gerdes (1987). Cardiac hypertrophy has been seen, rarely, in the AP rat. The macroscopic diagnosis of enlarged heart is a subjective view which depends on the skill of the prosector. In carcinogenicity studies the heart was only weighed in animals with macroscopic abnormalities. In the three AP rats where the macroscopic observation was found to be a real change, the heart weight was increased by 65 to 80 per cent and all three males had left ventricular hypertrophy and severe renal disease. Increased cardiac weights have been induced in the AP rat by administration of high doses of β blocking agents (Cruickshank *et al.*, 1984). In these animals there was no change in the

Table 6.1   Cardiac weights in the AP rat

Age (weeks)	Number/ sex	Mean heart weight			
		Weight (g)		Relative weight (% body weight)	
		Males	Females	Males	Females
12	10	1.25	0.91	0.36	0.40
34	20	1.80	1.18	0.29	0.36
58	25	1.72	1.14	0.23	0.29

myocardial cells, and the increased heart weight was considered to be an adaptive change to the increased workload of the heart. This type of adaptive change has been suggested for a variety of cardiac drugs (Whitehead *et al.*, 1979; Hoffman, 1984). This indicates that any factor which affects the haemodynamic status of the animal may affect cardiac function and weight.

Myocarditis

The most common lesions in the heart of the AP rat are inflammatory lesions. Pericarditis is a rare condition and is usually secondary to other infections within the thorax, including dosing accidents which perforate the oesophagus. Myocardial lesions, on the other hand, are common and occur in rats of all ages, with a higher incidence in males (Table 6.2). In animals less than 6 months of age the lesions are not visible macroscopically, and consist of very small foci of degenerate myofibres with a few mononuclear inflammatory

Table 6.2   Incidence of myocarditis in the AP rat

Duration of studies	Highest incidence of myocarditis observed in any study[a]	
	Males	Females
2 weeks	1/10 (10%)	1/10 (10%)
1 month	4/10 (40%)	2/10 (20%)
3 months	5/10 (50%)	2/10 (20%)
6 months	11/20 (55%)	5/20 (25%)
12 months	18/25 (72%)	8/25 (32%)
24 months	39/50 (78%)	21/50 (42%)

[a] Incidence derived from a database of 8880 control animals used in toxicology studies completed between 1960 and 1994.

cells, including the large cardiac histiocytes known as Anitschkow cells (Figure 47).

The foci may occur anywhere in the ventricular myocardium but are most common in the papillary muscle and subepicardial areas. In older animals fibrous replacement is the predominant feature (Figure 48) and inflammatory foci are infrequent. The areas of fibrosis increase in size with time indicating the progressive nature of the disease. The areas of subepicardial myocardium involved may be sufficiently large to be visible macroscopically as cream coloured areas on the surface of the heart (Figure 49). This lesion has been described as chronic progressive cardiomyopathy by Arceo *et al.* (1990). In the AP rat the incidence increases in studies up to 12 months in duration; thereafter the severity of the cardiomyopathy increases but there is only a slight increase in the numbers affected (Table 6.2). The incidence of myocarditis in the AP rat has increased since 1960; in 2 year studies, up to 1975, the incidence in males was always less than 50 per cent. The aetiology of the cardiomyopathy is not certain but it has been suggested that the foci of myocardial damage are due to ischaemia as a result of vascular disease (Ayers and Jones, 1978). This seems unlikely in the AP rat where animals in 2 week studies (aged about 8 weeks) have developed the disease, albeit usually of a minimal to mild severity. Few of the rats with cardiomyopathy show any disease in the coronary arteries although it has been suggested by Factor *et al.* (1984) that changes in the microvasculature are more important than coronary artery disease. It has also been suggested that there is a link between CPGN and cardiomyopathy (Van

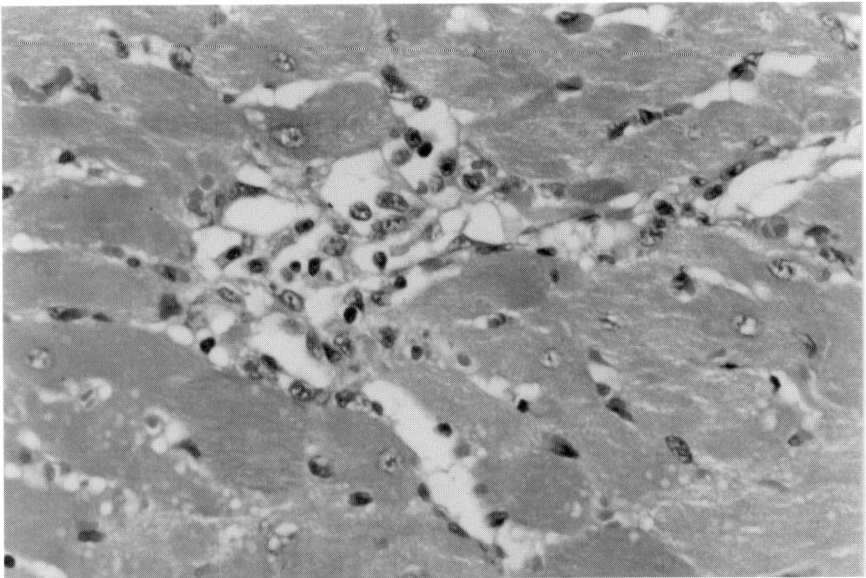

Figure 47 Focal chronic myocarditis in a male AP rat aged 3 months. The focus includes degenerate muscle fibres and a few inflammatory cells. ×128, H&E

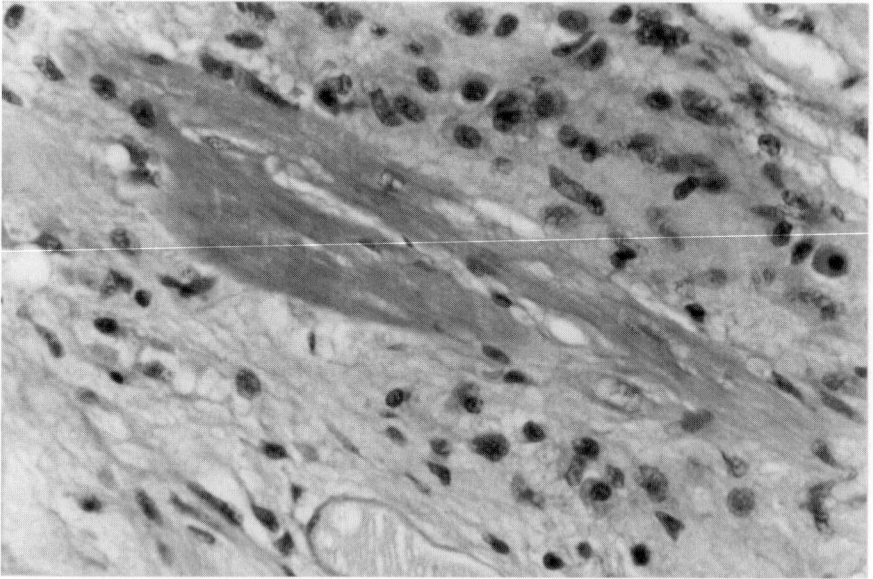

Figure 48 Chronic myocarditis showing fibrous replacement around degenerating muscle fibres. ×128, H&E

Figure 49 Myocardial fibrosis (↑). ×32, H&E

Vleet and Ferans, 1986). Both diseases are more common in males, but the myopathy is present in young animals with no overt signs of renal disease, although a mild proteinurea may be present. In the AP rat the incidence of cardiomyopathy has often been increased over control levels in the groups dosed with the test compound. This was observed with several drugs which affect cardiac function, including β adrenergic receptor blockers, antihypertensive agents and inotropic drugs. This might lend weight to the hypothesis that it is due to an effect upon the vascular system, but it has also been seen with drugs which have no known obvious effect on cardiac function. It is possible that a subclinical infection with an agent such as a virus is the cause, and that the incidence in groups treated with drugs having different pharmacological activities is related to an immunosupressant effect of the drugs. This would fit the pattern of the disease in the AP rat where some studies do not show any cardiac disease and others show a high incidence. Chronic progressive cardiomyopathy is common in F344 rats (Maeda *et al.*, 1985), SD rats (Lewis, 1994) and other Wistar strains (Maekawa *et al.*, 1983).

Mineralisation

Mineralisation of the myocardium is only seen in the dystrophic calcification secondary to renal disease. The mineralisation may be present as fine basophilic granules within the myocardial cells or in the media of the myocardial arterioles (Figure 50). Cartilaginous or osseous metaplasia of chordae

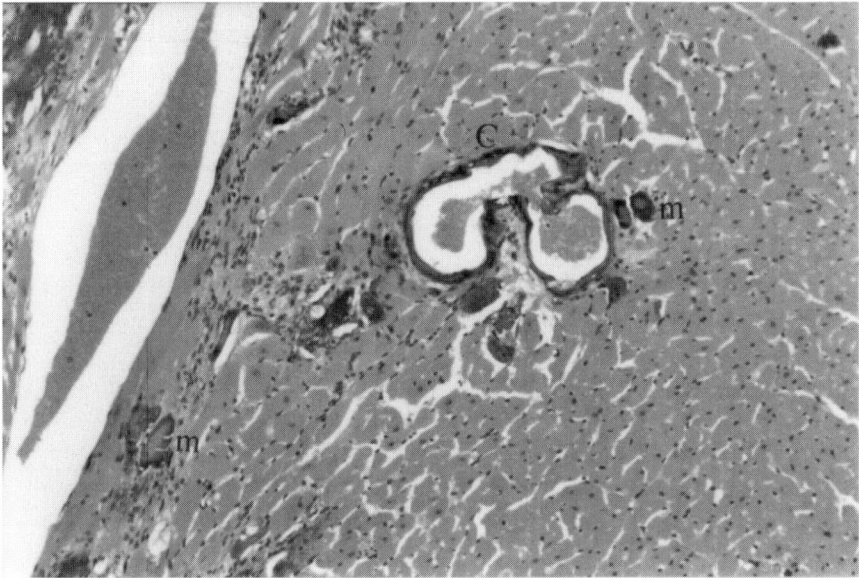

Figure 50   Mineralisation of myocardial cells (m) and coronary blood vessels (C). ×80, H&E

tendineae is a rare (<0.1 per cent) lesion (Figure 51) which does not appear to be secondary to any other condition. It may be a result of long term degenerative changes in the cordae caused by overstretching.

Valvular changes

Histological examination of the heart does not always include a section of a valve, but in approximately 90 per cent of animals the mitral valve is present in the section. This valve shows a myxomatous degeneration which increases in incidence and severity with age, from 1 to 24 months. It is more common in males and may reach a total incidence of 35 per cent, but in females the maximum level is 18 per cent. A similar high incidence has been seen in SD rats (Lewis, 1994). The cause of the valvular degeneration is not known but it has been reported in rats exposed to high altitude, castration and cold (Angrist *et al.*, 1960). None of these seems to be a likely cause of the spontaneous condition in the AP rat. Although the valvular degeneration may be extensive there is no evidence of the extra cardiac effects which occur in humans with mitral insufficiency caused by stenosis, i.e. left ventricular hypertrophy and dilatation, left atrial dilatation and passive congestion of the lungs. Since the valvular lesions are present in relatively young animals, although it progresses with age, it seems unlikely that the condition is merely a feature of aging. There does not appear to be an association with cardiomyopathy.

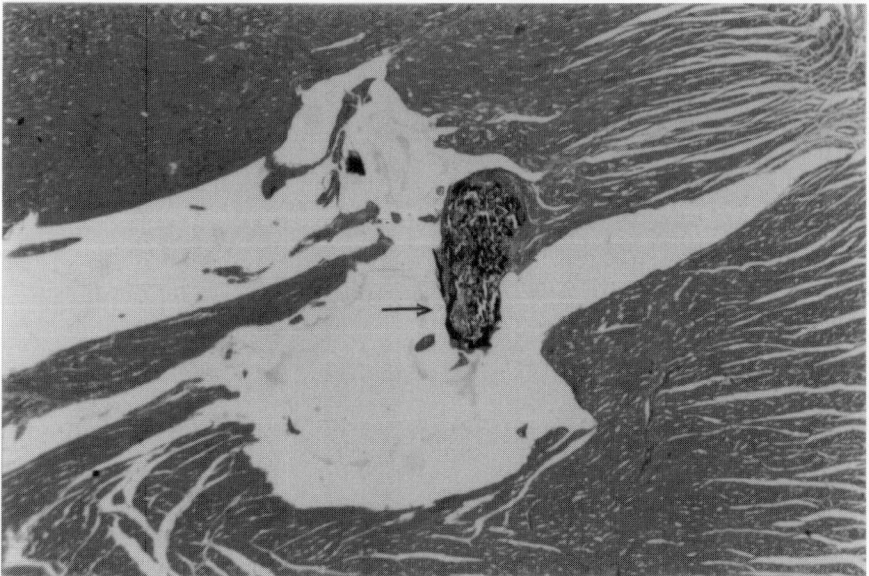

Figure 51   Osseous metaplasia (→) of chorda in left ventricle. ×32, H&E

Hypertrophy

Hypertrophy of the left ventricle has been seen in three male AP rats with severe renal disease. It has been induced in rats by administration of thyroxine (Zitnik and Roth, 1981) and an iron deficient diet (Neffgen and Korecky, 1972).

Thrombosis

Occasional thrombi (<1 per cent) have been seen in the left atrium. The atrium is grossly dilated and completely filled with thrombus, which in the AP rat has usually been an organised thrombus indicating that it has been present for some time. There has not been any associated cardiac pathology to account for the thrombosis in the AP rat, but it does occur more frequently in hypertensive rats (Wexler *et al.*, 1981), which suggests that hypertension may be a factor in the development of thrombosis. Kroes *et al.* (1981) reported an 8 per cent incidence of cardiac thrombosis in Wistar CPB rats, and Grice *et al.* (1969) reported it in rats given prolonged administration of cobalt.

Endocardial fibrosis

Subendocardial fibrosis, a proliferation of spindle cells between the endocardial endothelium and the cardiac muscle, is an uncommon condition seen only in AP rats over 18 months of age. The incidence in carcinogenicity studies has

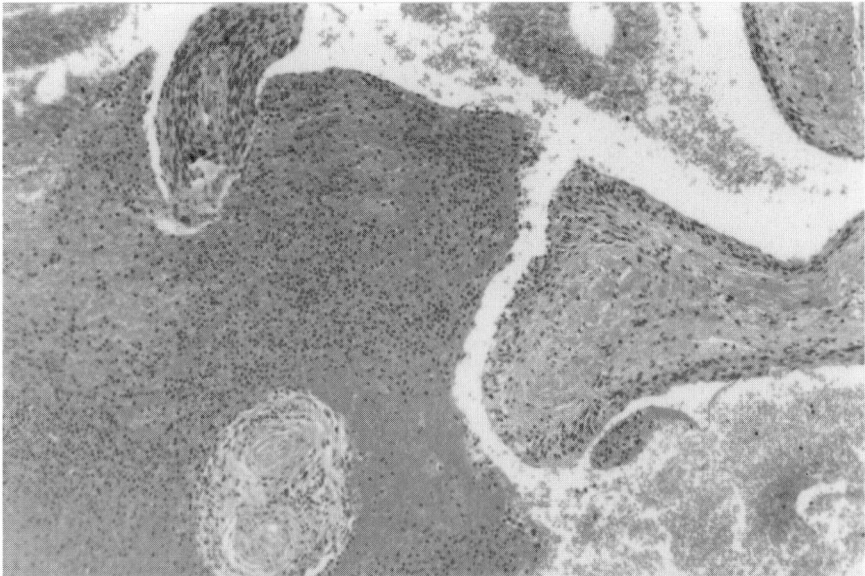

Figure 52   Subendocardial fibrosis of left ventricle and chordae tendineae. ×32, H&E

varied from 0 to 15 per cent. It is usually confined to the left ventricle and may also involve the chordae tendineae (Figure 52). The thickness of the proliferated cells is variable and mitotic figures are rare. Landes *et al.* (1988) have demonstrated positive staining for S-100 protein and suggest the Schwann cell is the most likely origin, but other workers have failed to demonstrate the antigen (Naylor *et al.*, 1986). Another factor is that the presence of S-100 protein is not diagnostic for Schwann cells, as it is present in other types of cell, and it has been shown that, ultrastructurally, the cells have no features which would indicate either Schwann cell or muscle cell origin.

6.1.2 Neoplastic Changes

Neoplasms of the heart are very rare. The database includes two tumours, diagnosed as fibrosarcomas, arising in the left ventricle. They were both in males which also had endocardial fibrosis (Figure 53). The heart has shown occasional metastatic infiltration by leukaemic cells. This low incidence of cardiac neoplasms is also reported for other strains such as the F344 (Alison *et al.*, 1987). Two strains of rats have a high spontaneous incidence of unusual tumours. The WAG/Rij (Wistar rat) has a 13 per cent incidence of hyperplasia/neoplasia of the aortic body (Van Zweiten *et al.*, 1979) and the NZR/Gd rat a 20 per cent incidence of atriocaval mesotheliomas in males (Goodall and Doesburg, 1981).

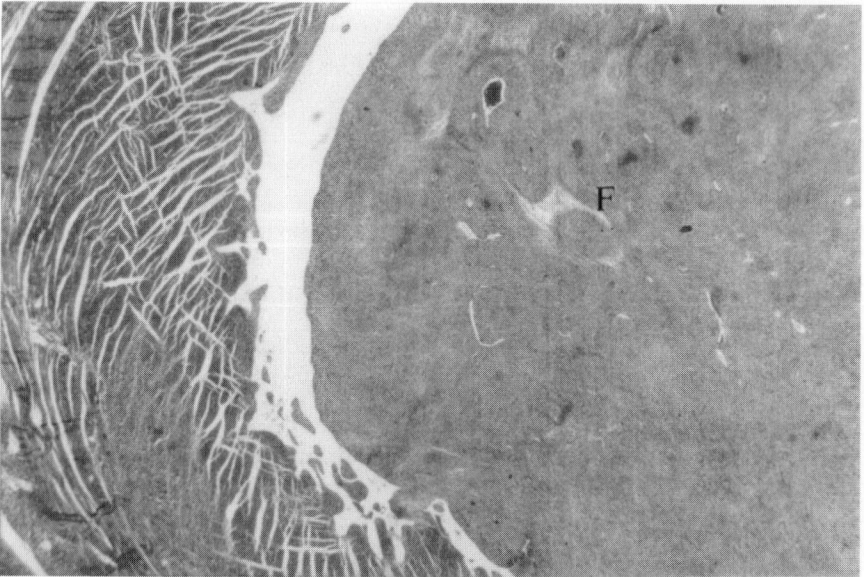

Figure 53 Fibrosarcoma (F) of the left ventricle in a male AP rat. ×8, H&E

6.2 Vascular System

6.2.1 *Non-neoplastic Diseases*

Polyarteritis

The most important vascular disease in the AP rat is a polyarteritis which was first described in the rat by Wilens and Sproul (1938). In the most severe form it can be recognised easily macroscopically, particularly in the mesenteric vessels, which are greatly enlarged, thickened and tortuous. The condition affects medium and large arterial branches but rarely arterioles. The earliest change is thought to be an inflammatory cell infiltration of the adventitia which spreads into the media with fibrinoid necrosis and disruption of the elastic and subintimal fibrin deposition (Figure 54). As the disease progresses, fibrosis replaces the necrosis, the vessels are greatly enlarged and may be totally occluded and often recanulated. It is most common in the mesenteric and testicular arteries, but other common sites include the kidney, heart, spleen, pancreas, gastro-intestinal tract and ovaries. It has never been seen in the aorta. The disease is more common in males, with incidence levels at 2 years ranging from 0 to 20 per cent in males and 0 to 8 per cent in females. This sex difference in incidence has been recorded for other strains (Goodman *et al.*, 1980; Richardson *et al.*, 1984). The aetiology is unknown but incidence levels can be modified by a variety of factors, including food restriction (Yu *et al.*, 1982). Aikawa and Kotetsky (1970) showed that arterial fibrinoid necrosis

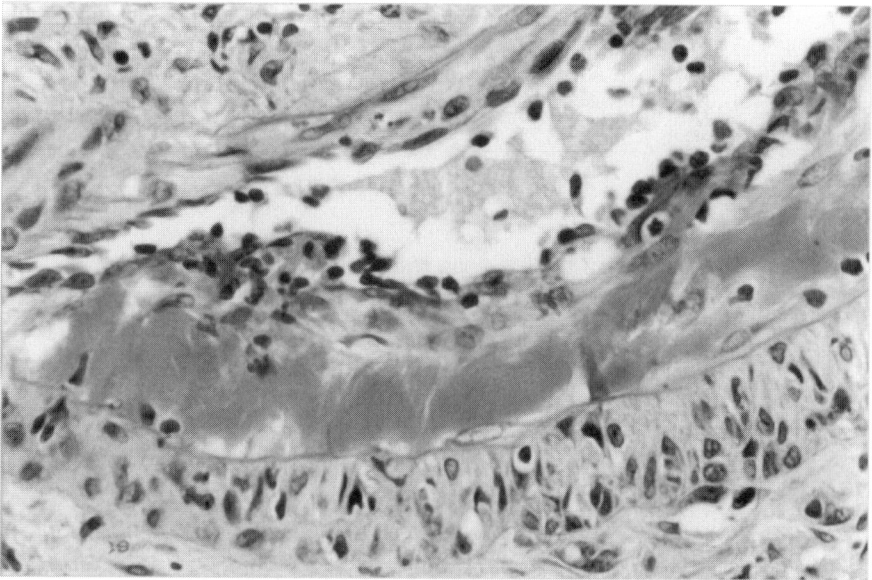

Figure 54   Polyarteritis in a male AP rat. ×128, H&E

105

developed in rats with severe renal hypertension of short duration and postulated that the polyarteritis-like disease is a result of prolonged renal hypertension. Administration of the drug bromocriptine, a prolactin inhibitor, to rats markedly reduced the incidence of renal disease and polyarteritis (Richardson *et al.*, 1984), and the dopamine agonist fenoldopam induced necrotising arteritis in the splanchnic vessels of SD rats (Kerns *et al.*, 1989). These findings lend weight to the theory that there is a link between renal disease and arteritis.

Mineralisation

Mineralisation of blood vessels occurs as a secondary condition in rats with severe kidney disease and parathyroid hyperplasia. It is common in the aorta which, macroscopically, is enlarged and rigid. Another type of aortic arteriosclerosis, seen in breeding female AP rats, is related to the number and frequency of litters. In this arterial mineralisation there is mucopolysaccharide accumulation in the tunica media followed by fibrosis, cartilaginous metaplasia and fibrosis (Tucker, 1971). This arteriosclerosis is thought to be related to the hormonal imbalances which occur in repeated pregnancies and lactation (Alper and Ruegner, 1969; Judd and Wexler, 1969). The involvement of lactation again indicates that the hormone prolactin may be involved in the aetiology of this disease.

Thrombosis and emboli

Thrombosis of pulmonary capillaries and small veins is a rare condition not associated with renal disease as has been reported for SD rats (Lewis, 1994). Hair emboli have been seen in pulmonary arteries in animals which have had intravenous injections of excipient. This has been reported by other workers (Innes *et al.*, 1958; Kast, 1985).

Hyperplasia

A few very old AP rats in the life-span study (all more than 36 months of age) showed vascular changes similar to hypertensive changes in man. Blood vessels in various organs, including kidneys and lungs, showed hyperplasia of medial and intimal layers. The changes were similar to those described in the arteries of spontaneously hypertensive rats (Limas *et al.*, 1980).

6.2.2 Neoplastic Diseases

Spontaneous vascular tumours are common in the AP rat and are described under the organs in which they have been found. The most common is the angioma of the lymph node, but angiomas and angiosarcomas have also been seen in the skin, muscle, liver and spleen. The angiomas are generally small tumours of vascular spaces lined by a single endothelial layer with little or no stroma. Angiosarcomas are locally invasive and poorly differentiated tumours, showing variable numbers of vascular spaces lined by pleomorphic multi-layered endothelial cells. Lung metastases were observed with one tumour. Spontaneous vascular neoplasms are uncommon in the F344 rat, where an incidence of 0.4 per cent of tumours has been observed (Mitsumori, 1990).

6.3 References

AIKAWA, M. and KOTETSKY, S. (1970) Arteriosclerosis of the mesenteric arteries of rats with renal hypertension. Electron microscopic observations, *American Journal of Pathology*, **61/3**, 293–304.

ALISON, R. H., ELWELL, M. R., JOKINEN, M. P., DITTRICH, K. L. and BOORMAN, G. A. (1987) Morphology and classification of 96 cardiac neoplasms in Fischer 344 rats, *Veterinary Pathology*, **24**, 488–94.

ALPER, R. and RUEGNER, W. R. (1969) Hormonal effects on the acid mucopolysaccharide composition of the rat aorta, *Journal of Atherosclerosis Research*, **10**, 19–25.

ANGRIST, A., OKA, M., NAKAO, K. and MARQUISS, J. (1960) Studies on experimental endocarditis: 1 Production of valvular lesions by mechanisms not involving infection or sensitivity factors, *American Journal of Pathology*, **36**, 181–99.

ARCEO, R. J., BISHOP, S. R., ELWELL, M. R., KERNS, W. D., MESFIN, G. M., RUBEN, Z., SANDUSKY, G. E. and VAN VLEET, J. F. (1990) Standard nomenclature of spontaneous pathological findings in the heart and vasculature of the laboratory rat. Initial Proposal Society of Toxicologic Pathology, in *Guides for Toxicological Pathology* STP/ARP/AFIP: Washington, DC.

AYERS, K. M. and JONES, S. R. (1978) The cardiovascular system, in BERNISCHKE, K., GARNER, F. M. and JONES, T. C. (Eds), *Pathology of Laboratory Animals*, Vol. 1, pp. 1–69, New York: Springer-Verlag.

CAMPBELL, S. E. and GERDES, A. M. (1987) Regional differences in myocyte dimensions and number in Sprague-Dawley rats from different suppliers, *Proceedings of the Society for Experimental Biology and Medicine*, **186**, 221–7.

CRUICKSHANK, J. M., FITZGERALD, J. D. and TUCKER, M. J. (1984) Beta-adrenoreceptor blocking drugs, pronethalol, propranalol and practolol, in LAWRENCE, D. R., MCLEAN, A. E. M. and WEATHERAL, M. (Eds), *Safety Testing of New Drugs, Laboratory Predictions and Clinical Performance*, pp. 93–123, London: Academic Press.

FACTOR, S. M., MINASE, T., CHO, S., FEIN, F., CAPASSO, J. M. and SONNENBLICK, E. H. (1984) Coronary microvascular abnormalities in the hypertensive-diabetic rat. A primary cause of cardiomyopathy? *American Journal of Pathology*, **116**, 9–20.

GOODALL, C. M. and DOESBURG, R. M. N. (1981) Age-specific incidence of neoplasms in untreated NZR/Gd inbred rats: an inbred strain with cardio-vascular tumours and liver glycogen storage disease, *Journal of Pathology*, **135**, 147–57.

GOODMAN, D. G., WARD, J. M., SQUIRE, R. A., PAXTON, M. B., REICHARDT, W. D., CHU, K. C. and LINHART, M. S. (1980) Neoplastic and non-neoplastic lesions in ageing Osborne-Mendel rats, *Toxicology and Applied Pharmacology*, **55**, 433–47.

GRICE, H. C., MUNRO, I. C. and WIBERG, G. S. (1969) The pathology of experimentally induced cobalt cardiomyopathies. A comparison with beer drinkers cardiomyopathy, *Clinical Toxicology*, **2**, 273–7.

HOFFMAN, K. (1984) Toxicological studies with nitrendipene, in SCRIABINE, A., VANOV, S. and DECK, K. (Eds), *Nitrendipine*, pp. 25–31, Baltimore: Urban and Schwarzenberg.

INNES, J. R. M., DONATIE, E. J. and YEVICH, P. P. (1958) Pulmonary lesions in mice due to fragments of hair, epidermis and extraneous matter accidentally injected in toxicity experiments, *American Journal of Pathology*, **34**, 161–7.

JUDD, D. and WEXLER, B. C. (1969) The role of lactation and weaning in the pathogenesis of arteriosclerosis in female breeder rats, *Journal of Atherosclerosis Research*, **10**, 435–9.

KAST, A. (1985) Pulmonary hair embolism, rat, in JONES, T. C., MOHR, U., HUNT, R. D. (Eds), *Pathology of Laboratory Animals*, pp. 186–94, Berlin and New York: Springer-Verlag.

KERNS, W. D., ARENA, E., MACIA, R. A., BUGELSKI, P. J., MATHEWS, W. D. and MORGAN, D. G. (1989) Pathogenesis of arterial lesions produced by dopaminergic compounds in the rat, *Toxicologic Pathology*, **17**, 203–13.

KROES, R., GARBIS-BERKVENS, J. M., DE VRIES, T. and VAN NESSELROOY, J. H. J. (1981) Histopathological profile of a Wistar rat stock including a survey of the literature, *Journal of Gerontology*, **36/3**, 259–79.

LANDES, C. H., RUEFENACHT, H. J., NAYLOR, D. C. and KRINKE, G. J. (1988) Rat endomyocardial disease: a neural origin? *Experimental Pathology*, **34**, 65–9.

LEWIS, D. J. (1994) Non-neoplastic lesions in the cardiovascular system, in MOHR, U., DUNGWORTH, D. L. and CAPEN, C. C. (Eds), *Pathobiology of the Aging Rat*, Vol. 1, pp. 300–9, Washington: ILSI Press.

LIMAS, C., WESTRUM, B. and LIMAS, C. J. (1980) The evolution of vascular changes in the spontaneously hypertensive rat, *American Journal of Pathology*, **98**, 357–84.

MAEDA, H., GLEISER, C. A., MASORO, E. J., MURATO, I., MCMAHAN, C. A. and YU, B. P (1985) Nutritional influences on aging Fischer 344 rats: pathology, *Journal of Gerontology*, **40/6**, 671–88.

MAEKAWA, A., ONODERA, H., TANIGAWA, H., FURUTA, K., KODOMA, Y., HORIUCHI, S. and HAYASHI, Y. (1983) Neoplastic and non-neoplastic lesions in aging Slc Wistar rats, *Journal of Toxicological Science*, **8**, 279–90.

MITSUMORI, K. (1990) Blood and lymphatic vessels, in BOORMAN, G. A., EUSTIS, S. L., ELWELL, M. R., MONTGOMERY, C. A. and MACKENZIE, W. E. (Eds), *Pathology of the Fischer Rat, Reference and Atlas*, pp. 473–84, San Diego: Academic Press.

NAYLOR, D. C., KRINKE, G. and ZAK, F. (1986) A comparison of endomyocardial disease in the rat with endomyocardial fibrosis in man, *Journal of Comparative Pathology*, **96**, 473–83.

NEFFGEN, J. F. and KORECKY, B. (1972) Cellular hyperplasia and hypertrophy in cardiomegalies induced by anaemia in young and adult rats, *Circulation Research*, **30**, 104–13.

108

RICHARDSON, B. P., TURKALJ, I. and FLUCKINGER, E. (1984) Bromocriptine, in LAURENCE, D. R., MCLEAN, A. E. M. and WETHERALL, M. (Eds), *Safety Testing of New Drugs,* Laboratory Predictions and Clinical Performance, pp. 19–63, London: Academic Press.

TANASE, H., YAMORI, Y., HANSEN, C. T. and LOVENBERG, W. (1982) Heart size in inbred strains of rats. Part 1: genetic determination of the development of cardiovascular enlargement in rats, *Hypertension,* **4**, 864–72.

TUCKER, M. J. (1971) Effect of clofibrate on spontaneous arteriosclerosis in rats, *Atherosclerosis,* **13**, 255–65.

VAN VLEET, J. F. and FERRANS, V. J. (1986) Myocardial diseases of animals, *American Journal of Pathology,* **124**, 98–178.

VAN ZWEITEN, M. J., BUREK, J. D., ZURCHER, C. and HOLLANDER, C. F. (1979) Aortic body tumours and hyperpasia in the rat, *Journal of Pathology,* **128**, 99–112.

WEXLER, B. C., MCMURTRY, J. P. and IAMS, S. G. (1981) Histopathological changes in aging male vs female spontaneously hypertensive rats, *Journal of Gerontology,* **36**, 514–19.

WHITEHEAD, P. N., CHESTERMAN, H., STREET, A. E., PRENTICE, D. E., HEYWOOD, R. and SADO, T. (1979) Toxicity of nicarpidine hydrochloride, a new vasodilator, in the Beagle dog, *Toxicology Letters,* **4**, 57–9.

WILENS, S. L. and SPROUL, E. E. (1938) Spontaneous cardiovascular disease in the rat, *American Journal of Pathology,* **14**, 177–83.

YU, B. P., MASORO, E. J., MURATA, I., BERTRAND, H. A. and LYND, F. T. (1982) Life span study of SPF Fischer 344 male rats fed ad libitum or restricted diets. Longevity, growth, lean body mass and disease, *Journal of Gerontology,* **37**, 130–41.

ZITNIK, G. and ROTH, G. S. (1981) Effects of thyroid hormones on cardiac hypertrophy and adrenergic receptors during ageing, *Mechanisms of Ageing and Development,* **15**, 19–28.

7

The Respiratory System

7.1 Nasal Cavities

7.1.1 Non-neoplastic Changes

The nasal cavities are inspected at necropsy but have not been routinely taken for histological examination in toxicology studies in the AP rat, except in inhalation studies and in one carcinogenicity study. In these studies a transverse section was taken through the nasoturbinates (Figure 55). The only non-neoplastic changes seen were a mild rhinitis with some exudate into the nasal cavities in five animals and the presence of eosinophilic globules in the olfactory epithelium of 2–3 per cent of animals in the 2 year study. This latter change and other non-neoplastic conditions have been described by St. Clair and Morgan (1992).

7.1.2 Neoplastic Changes

No spontaneous nasal tumours have been seen in the AP rat but, as the nasal cavities are not examined in regulatory toxicology studies, this cannot be considered an accurate figure, although all rat strains show a low incidence of nasal tumours (Boorman *et al.*, 1990; Maronpot, 1990; Monticello *et al.*, 1990). The few tumours observed include polypoid adenomas, adenocarcinomas and squamous carcinomas. Nasal tumours have been induced by a carcinogen, quinoxaline 1,4 dioxide (Tucker, 1975).

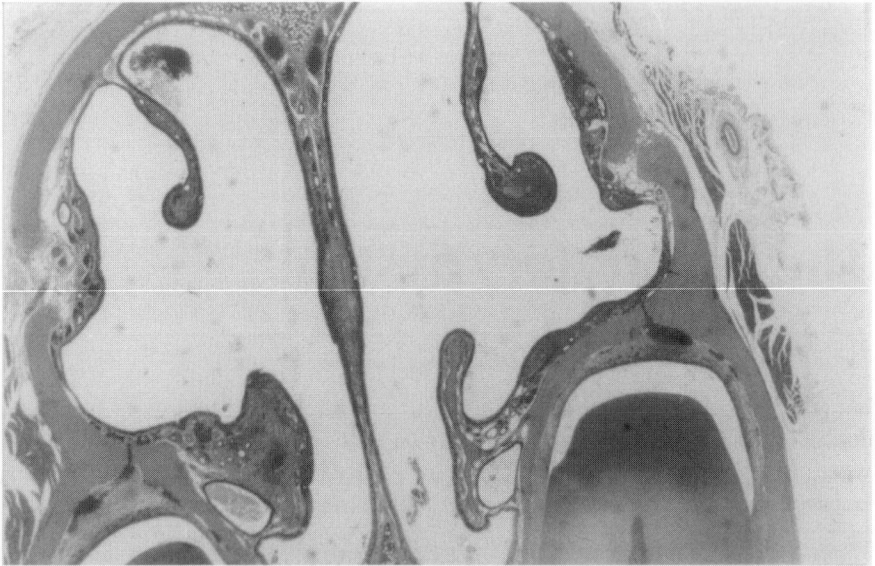

Figure 55   Transverse section through the nasal cavities of an AP rat. ×8, H&E

7.2   Larynx and Trachea

7.2.1   *Non-neoplastic Changes*

The larynx has also not been examined routinely and has only been seen occasionally in tracheal sections. No abnormalities have been observed in the AP rat and none was seen in 3548 F344 rats reviewed by Goodman *et al.* (1979). The trachea is examined in all animals and has shown few lesions in the AP rat. Approximately 10 per cent have shown a mild chronic tracheitis with some dilatation of subepithelial glands.

7.2.2   *Neoplastic Changes*

No tumours have been observed in the AP rat but Goodman *et al.* (1979) reported a single occurrence each of carcinoma *in situ*, squamous cell carcinoma, adenocarcinoma and fibroma in the trachea in their review of F344 rats.

7.3   Lungs

7.3.1   *Non-neoplastic Findings*

The method of fixation of the lungs is by immersion except for inhalation studies where the lungs were perfused fixed. For twenty years the whole lung was sectioned for histological examination but recently this has been reduced to macroscopic examination of the whole lung and histological examination of two lobes and the major bronchi.

Pneumonia

The advent of SPF animals has greatly reduced the incidence of infectious disease in the lung. When the AP rat was bred in conventional laboratory conditions bronchopneumonia was a major cause of death (Paget and Lemon, 1965). Lung infections have only been an important cause of morbidity and mortality when the AP rat colony was infected with *Pasteurella pneumotropica* during the 1960s. The bronchopneumonia was characterised by bronchiectatic abscesses, which produced clinical signs of respiratory distress with rapid, shallow breathing and wheezing in affected animals. When an outbreak of *Pasteurella* pneumonia occurred in a toxicology study the animals were treated with antibiotics to prevent the usual high mortality which would necessitate termination of the study, a practice which would not be considered acceptable in modern toxicological work.

Lobar pneumonia has been seen as a result of misdosing, or as a terminal condition; histologically, one or more lobes are consolidated by fluid and inflammatory cells in the alveoli without involvement of the airways. Infection of the colony with Sendai virus in the 1970s did not produce significant lung pathology in the adult rat. Weanling animals showed a typical viral pneumonia with bronchiolitis and interstitial inflammation. The disease had a high mortality in weanlings but, in the adult animals, showed only an increase in lymphoid tissue in the lung and foci of chronic alveolitis. Since the colony has been vaccinated against Sendai, inflammatory lung conditions have only been seen sporadically in the AP rat, and no specific pathogens have been identified.

Aspiration pneumonia occurs occasionally in old rats as a result of aspiration of food particles, which can usually be identified within the area of inflammation. One of the causes is thought to be the pressure effect of large pituitary tumours on the brain and swallowing reflexes (Dixon and Jure, 1988).

Pleuritis

Pleuritis has only been seen secondary to other inflammatory conditions in the thorax, chiefly after dosing accidents. The visceral pleura shows chronic or acute inflammatory cell infiltration and fibrosis with adhesions between the lobes, with the pericardium, or the wall of the thoracic cavity. Finger-like proliferation of the pleural mesothelium has been seen as a reaction to neoplasms within the lung.

Oedema/haemorrhage/congestion

These are all relatively uncommon conditions in the AP rat and are mostly seen in older animals. Pulmonary oedema is seen as a secondary condition to cardiac and renal disease, and also occurs as an agonal change. Haemorrhage may also be an agonal or a terminal event, secondary to neoplasms in the lung. Congestion of the lungs is a frequent finding in sudden death or in animals which are killed by inhalation of gases such as carbon dioxide.

Alveolar macrophages

Alveolar macrophages are a common finding in AP rats of all ages. The incidence is variable and ranges from 0 to a maximum of 30 per cent in both sexes. They are most frequent in the subpleural alveoli and are found less frequently near major airways of blood vessels (Figure 56). When large, the foci are visible macroscopically as greyish-white areas of the lung surface. The macrophages are large with foamy cytoplasm which contains phospholipids, cholesterol and free fatty acids, and they may be associated with a small number of inflammatory cells (Figure 57). In old animals where the lesions have been present for some time there is metaplasia of the alveoli by type II epithelial cells (Figure 58). The macrophages may break down finally with the formation of cholesterol crystals, and this may provoke a foreign-body reaction with giant cells and fibrosis. Alveolar macrophages have been reported in other strains such as the F344 (Shibuya *et al.*, 1986) and other Wistar rats (Dungworth *et al.*, 1992). A variety of factors may be involved in the aetiology of alveolar macrophage infiltration. Diets deficient in essential fatty acids increase the incidence (Bernick and Alfin-Slater, 1963) and, as pulmonary macrophages are more common in non-SPF animals, pulmonary inflammatory processes are

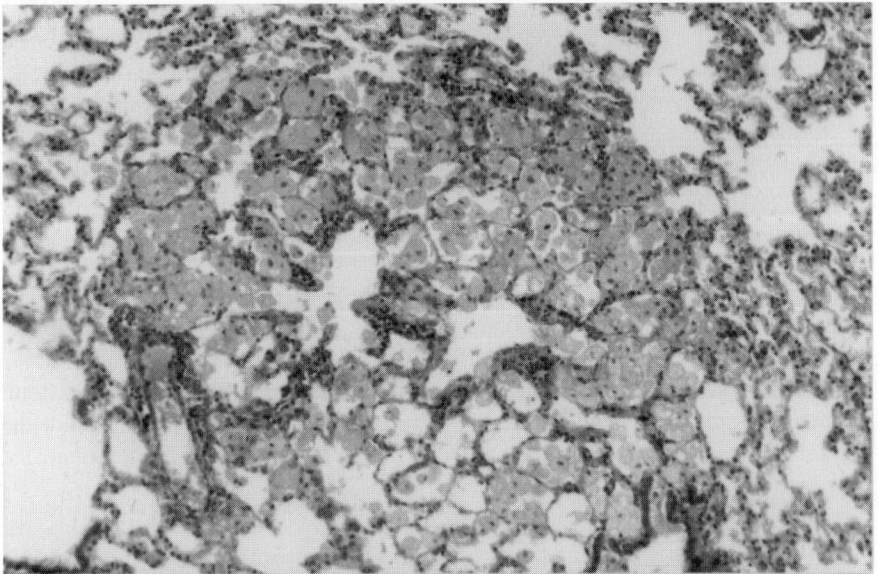

Figure 56   Lung of male AP rat showing a focus of alveolar macrophages. ×8, H&E

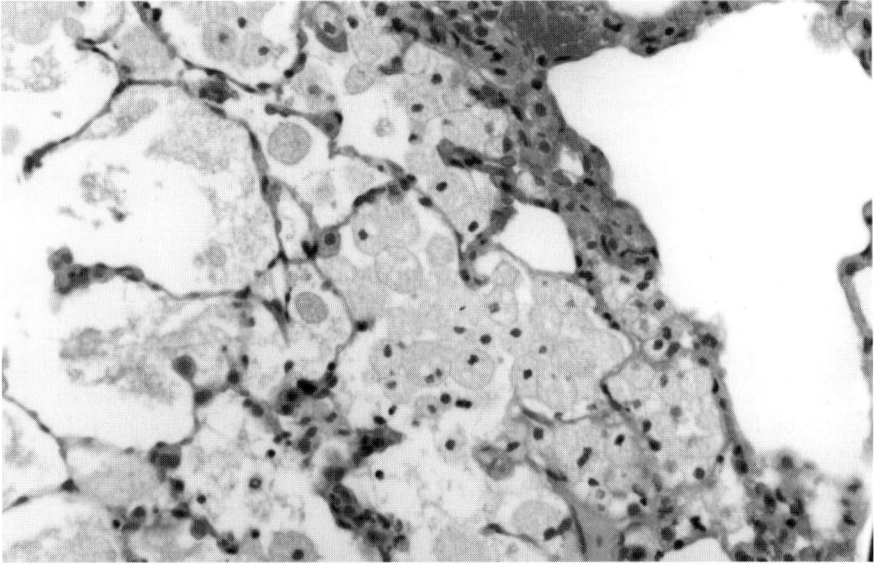

Figure 57 Large foamy alveolar macrophages in the lung of a male AP rat. ×128, H&E

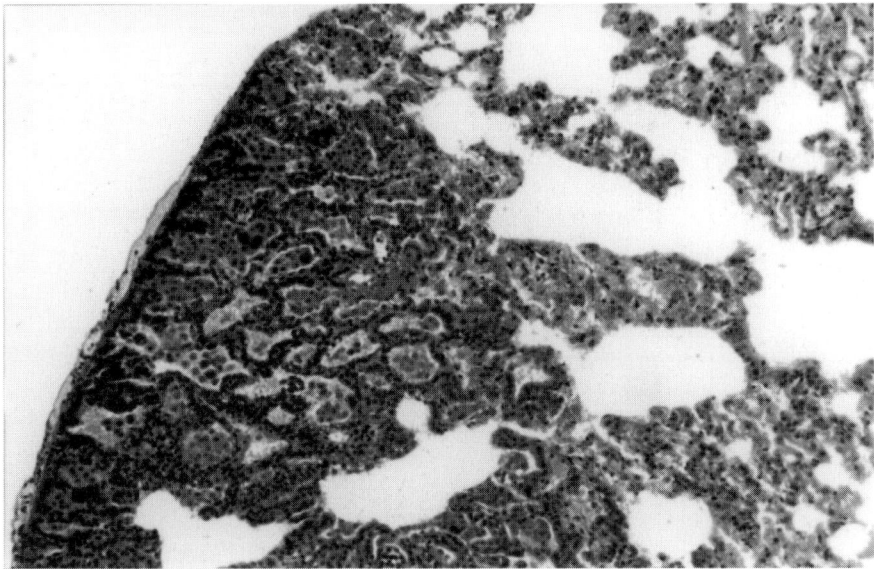

Figure 58 Long-standing focus of alveolar macrophages with metaplasia of type II alveolar cells. ×32, H&E

also thought to be involved. A marked increase in alveolar macrophages in the lungs of AP rats has been induced by a variety of cationic and amphiphilic drugs, and this has been reported with other strains (Lüllman *et al.*, 1978; Reasor, 1981).

Pigmented macrophages are quite common in old (>30 months) AP rats and they stain for iron. Minor haemorrhage is the most probable cause as they are increased greatly in lung tissue where there is severe haemorrhage related to a neoplasm.

Bronchial associated lymphoid tissue

Bronchial associated lymphoid tissue (BALT) is not a conspicuous feature of the lungs of AP rats, except for the occasion when the colony was infected with *Pasteurella*; infected animals showed very extensive BALT similar to that described for the conventionally reared colony by Paget and Lemon (1965). In the virus infection with Sendai, small perivascular foci of lymphocytes were the most common feature.

Alveolar hyperplasia

Alveolar hyperplasia is the condition when the alveoli are lined by bronchiolar cells (Clara cells) or alveoli type II cells. It is rare in AP rats less than 24 months old, but is not uncommon in animals over 30 months of age. The

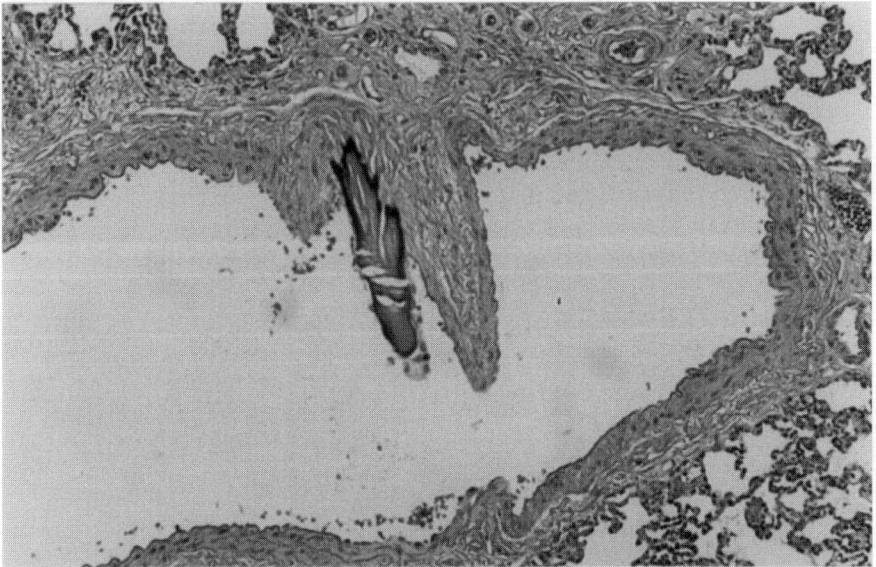

Figure 59 Focal subintimal calcification in a pulmonary vessel of a 3 month old AP rat. ×32, H&E

hyperplasia does occur in association with alveolar macrophages but is also seen without any other evidence of pulmonary changes. Similar low incidences have been reported for the F344 (Boorman, 1985) and Wistar rats (Dungworth *et al.*, 1992).

Vascular lesions

Vascular lesions in the lung include hair embolisms in pulmonary vessels. A very common lesion (100 per cent incidence), which does not appear to be age related, is focal subintimal calcification (Figures 59 and 60). The pathogenesis and aetiology of this lesion are not known but it is reported in other Wistar rats (Dungworth *et al.*, 1992). Medial hypertrophy does occur rarely in the old AP rat but must be distinguished from the thickened arteries seen quite frequently in young animals; in this condition the appearance of greatly thickened arterial walls is due to oblique sectioning of the thick-walled pulmonary vessels (Meyrick *et al.*, 1978).

Mineralisation

Alveolar calcification occurs in renal secondary hyperparathyroidism (Figures 61 and 62) and may be small or large deposits. Small gobbets of bone are seen occasionally in the alveoli; Innes *et al.* (1956) thought that they were inhaled particles of fish meal from the diet. They have, however, been observed in AP

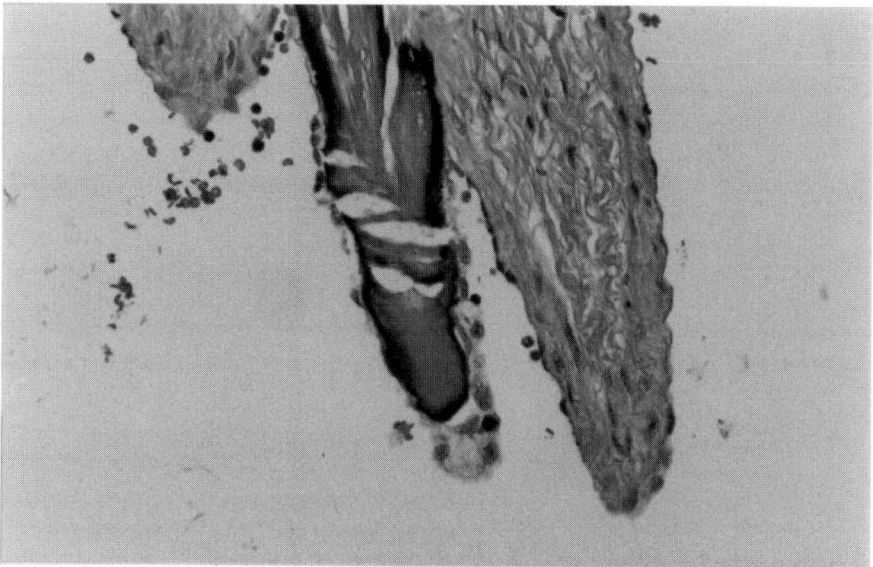

Figure 60 A higher power view of focal subintimal calcification. There is no reaction to the calcified material. ×128, H&E

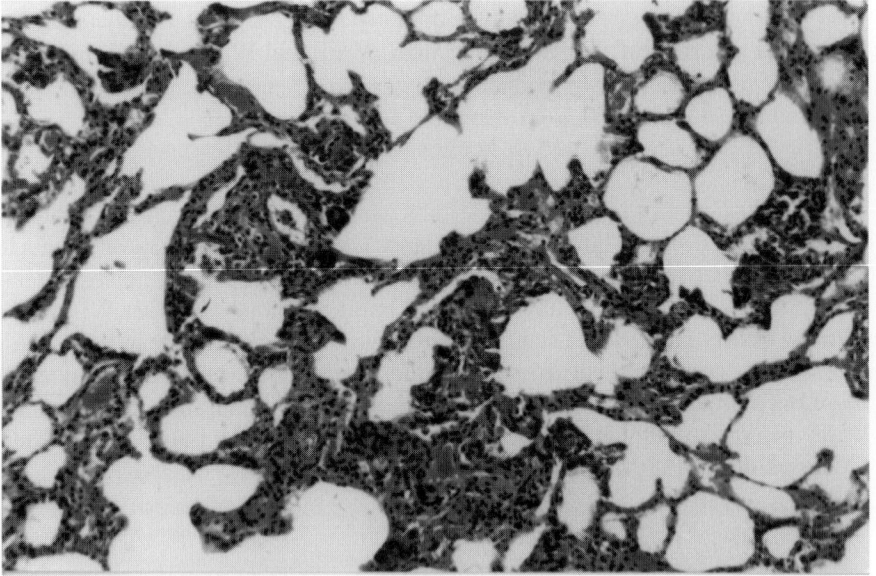

Figure 61 Area of mineralisation in lung of a male AP rat with renal secondary hyperparathyroidism. ×80, H&E

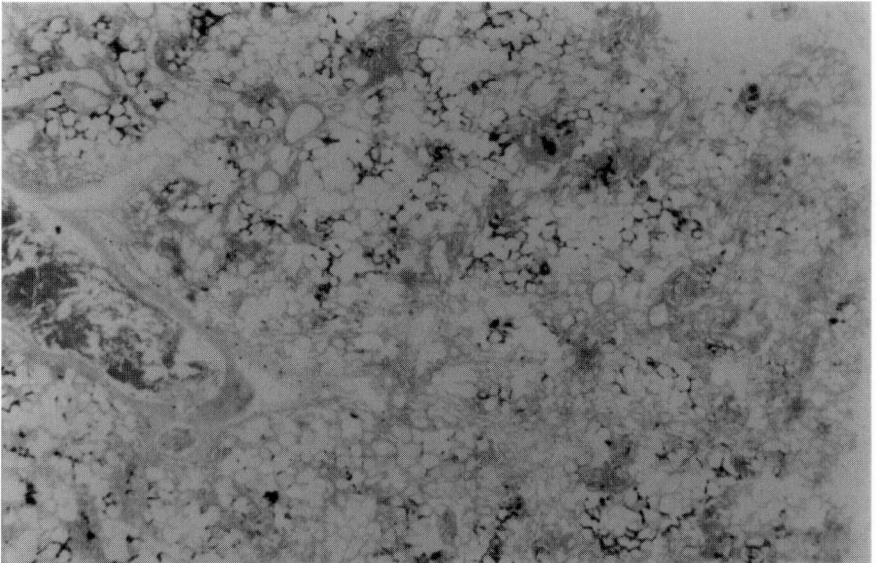

Figure 62 Multifocal calcification of the lung in a male AP rat with renal secondary hyperparathyroidism. ×32, Von Kossa

rats since fish meal was removed from the diet, so that they are more likely to represent small foci of osseous metaplasia of unknown histogenesis. Inflammatory conditions related to aspiration of food or other foreign material may produce a foreign body giant cell reaction. The cellulose coat of plant material can usually be identified in these granulomas.

7.3.2 Neoplastic changes

Primary pulmonary tumours are rare in the AP rat, with the maximum incidence in any control group being 1 per cent, and this has not changed since 1960. The histological types which have been seen in the AP rat are shown in Table 7.1. The youngest animal with a lung tumour was a male aged 18 months. This low incidence is reported in all other strains, including the F344 (Sollveld *et al.*, 1984; Haseman *et al.*, 1984), SD (Mori and Fujii, 1973; Stula, 1975), Osborne-Mendel (Goodman *et al.*, 1980) and Wistar (Kroes *et al.*, 1981).

Epithelial tumours

The most common epithelial tumour in the AP rat is the bronchio-alveolar adenocarcinoma. The tumours were all large, occupying the whole of a lobe with metastases to other lobes in two animals. Mild respiratory distress had been noted in two of these animals. The tumours had an irregular border and showed a mixture of glandular and papillary formation and a high mitotic rate. Adenomas are less frequent and tend to be smaller with a distinct border and a well-differentiated pattern. As there is a progression from adenoma to adenocarcinoma, there are tumours which are borderline between the two.

Table 7.1 Incidence of lung tumours in the AP rat

Type	Number observed[a]	
	Male	Female
Broncho-alveolar adenoma	3	1
Broncho-alvoelar adenocarcinoma	3	4
Squamous carcinoma	3	0
Lymphosarcoma	2	0
Fibrosarcoma	3	0
Liposarcoma	1	0

[a] Number of tumours observed in a database of 8880 AP rats (including 2800 males and 2500 females in 2 year studies) used in toxicology studies between 1960 and 1992. Highest incidence observed in any study was 1%.

Squamous cell carcinomas have been found and these showed the histological characteristics of the carcinoma in the skin, but keratin and pearls were sparse. The major part of the tumours was composed of cells showing dysplasia, atypia and a high mitotic rate. The origin of the squamous cell is uncertain, but squamous metaplasia occurs in bronchiolar epithelium prior to the development of squamous carcinomas induced by radiation and chemicals (Kuschner and Laskin, 1970; Reznik-Schüller and Gregg, 1981). Alveolar type II cells may also undergo squamous metaplasia (Adamson and Bowden, 1979). This would suggest that whatever the cell of origin it is likely to have undergone squamous metaplasia prior to tumour development.

Mesenchymal tumours

Only two mesenchymal tumours have been observed: a fibrosarcoma in a male and a liposarcoma in a female. In the non-SPF AP rat the most common pulmonary tumour (3.2 per cent) was a primary lymphosarcoma; these tumours were considered to have arisen in the extensive BALT present in the lungs (Paget and Lemon, 1965). Two lymphosarcomas in the SPF AP rat were found in animals which had a severe infection with *Pasteurella pneumotropica* and an associated extensive BALT. These tumours were confined to the lung, and no primary lymphosarcomas of the lung have been seen since that outbreak of infection ended in the 1960s. The neoplastic cells replaced the BALT and extended deep into the lung parenchyma. They were mixed lymphocytic/lymphoblastic tumours and in most cases invaded the epithelium and airways of the bronchi. Bronchial adenomas have been induced in the AP rat by 2-acetylaminofluorene (unpublished observations).

The lung is the most common site for secondary tumours although, in general, metastases are not common (maximum 3 per cent). Lung metastases have been found for osteosarcoma, hepatocellular carcinoma, phaeochromocytoma, adreno-cortical carcinoma, islet cell carcinoma, mammary adenocarcinoma, thyroid 'C' cell carcinomas, skin squamous carcinomas, uterine adenocarcinomas, leukaemias and other lymphomas. They are also uncommon in other strains except for a high incidence of uterine adenocarcinoma metastases in the Han:Wistar rat strain (Deerberg *et al.*, 1981).

7.4 References

ADAMSON, I. Y. R. and BOWDEN, D. H. (1979) Bleomycin-induced injury and metaplasia of alveolar type 2 cells, *American Journal of Pathology*, **96**, 531–44.
BERNICK, S. and ALFIN-SLATER, R. B. (1963) Pulmonary infiltration of lipid in essential fatty acid deficiency, *Archives of Pathology*, **75**, 13–20.
BOORMAN, G. A. (1985) Bronchiolar/alveolar hyperplasia, lung, rat, in JONES, T. C., MOHR, U. and HUNT, R. D. (Eds), *Pathology of Laboratory Animals*, pp. 177–9, Berlin and New York: Springer.

BOORMAN, G. A., MORGAN, K. T. and URAIH, L. C. (1990) Nose, larynx and trachea, in BOORMAN, G. A., EUSTIS, S. L., ELWELL, M. R., MONTGOMERY, C. A. and MACKENZIE, W. F. (Eds), *Pathology of the Fischer Rat*, pp. 315–37, New York: Academic Press.

DEERBERG, F., RAPP, K. G., PITTERMANN, W. and REHM, S. (1981) Uncommon frequency of adenocarcinomas of the uterus in virgin Han:Wistar rats, *Veterinary Pathology*, **18**, 707–813.

DIXON, D. and JURE, M. N. (1988) Diagnostic exercise, Pneumonia in a rat, *Laboratory Animal Science*, **38**, 727–8.

DUNGWORTH, D. L., ERNST, H., NOLTE, T. and MOHR, U. (1992) Non-neoplastic lesions in the lungs, in MOHR, U., DUNGWORTH, D. L. and CAPEN, C. C. (Eds), *Pathobiology of the Aging Rat*, Vol. 1, pp. 141–60, Washington: ILSI Press.

GOODMAN, D. G., WARD, J. M., SQUIRE, R. A., CHU, K. C. and LINHART, M. S. (1979) Neoplastic and non-neoplastic lesions in aging F344 rats, *Toxicology and Applied Pharmacology*, 48, 237–48.

GOODMAN, D. G., WARD, J. M., SQUIRE, R. A., PAXTON, M. B., REICHARDT, W. D., CHU, K. C. and LINHART, M. S. (1980) Neoplastic and non-neoplastic lesions in aging Osborne-Mendel rats, *Toxicology and Applied Pharmacology*, **55**, 433–47.

HASEMAN, J. K., HUFF, J. E. and BOORMAN, G. A. (1984) Use of historical control data in carcinogenicity studies in rodents, *Toxicologic Pathology*, **12**, 126–35.

INNES, J. R. M., YEVICH, P. P. and DONATI, E. J. (1956) Note on origin of some fragments of bone in lungs of laboratory animals, *Archives of Pathology*, **61**, 401–6.

KROES, R., GARBIS-BERKVENS, J. M., DE VRIES, T. and VAN NESSELROOY, J. H. J. (1981) Histopathological profile of a Wistar rat stock including a survey of the literature, *Journal of Gerontology*, **36**, 259–79.

KUSCHNER, M. and LASKIN, S. (1970) Pulmonary epithelial tumors and tumor-like proliferations in the rat, in NETTESHEIM, P., HANNA, M. G. and DEATHERAGE, J. W. (Eds), *Morphology of Experimental Respiratory Carcinogenesis*, pp. 203–26, Oak Ridge, Tennessee: USAEC.

LULLMANN, H. R., LULLMANN-RAUCH, R. and WASSERMANN, O. (1978) Lipidosis induced by amphiphilic cationic drugs, *Biochemical Pharmacology*, **27**, 1103–8.

MARONPOT, R. R. (1990) Pathology working group review of selected upper respiratory tract lesions in rats and mice, *Environmental Health Perspectives*, **85**, 331–52.

MEYRICK, B., HISLOP, A. and REID, L. (1978) Pulmonary arteries of the normal rat: the thick walled oblique muscle segment, *Journal of Anatomy*, **125**, 209–21.

MONTICELLO, T. M., MORGAN, K. T., URIAH, L. (1990) Non-neoplastic lesions, nasal lesions in rats and mice, *Environmental Health Perspectives*, **85**, 249–74.

MORI, S. and FUJII, T. (1973) Spontaneous tumours in Sprague-Dawley JCL rats, *Experimental Animals* (Tokyo), **22**, 127–38.

PAGET, G. E. and LEMON, P. G. (1965) The interpretation of pathological data, in RIBELIN, W. E. and MCCOY, J. R. (Eds), *The Pathology of Laboratory Animals*, pp. 382–405, Springfield, Illinois: Charles C. Thomas.

REASOR, M. J. (1981) Drug induced lipidosis and the alveolar macrophage, *Toxicology*, **20**, 1–33.

REZNIK-SCHULLER, H. M. and GREGG, M. (1981) Pathogenesis of lung tumours induced by N-nitrosoheptamethyleneimine in the rat, in REZNIK-SCHULLER, H. M. (Ed.), *Comparative Respiratory Tract Carcinogenesis*, pp. 95–116, Boca Raton, Florida: CRC Press.

SHIBUYA, K., TAJIMA, M., YAMATE, J., SUTOH, M. and KUDOW, S. (1986) Spontaneous occurrence of pulmonary foam cells in Fischer 344 rats, *Japanese Journal of Veterinary Science*, **48**, 413–17.

SOLLVELD, H. A., HASEMAN, J. K. and MCCONNELL, E. E. (1984) Natural history of body weight gain, survival and neoplasia in the F344 rat, *Journal of the National Cancer Institute*, **72**, 929–40.

ST. CLAIR, M. B. G. and MORGAN, K. T. (1992) Changes in the upper respiratory tract, in MOHR, U., DUNGWORTH, D. L. and CAPEN, C. C. (Eds), *Pathobiology of the Aging Rat*, pp. 111–27, Washington: ILSI Press.

STULA, E. F. (1975) Naturally occurring pulmonary tumors of epithelial origin in Charles River-CD rats, *Bulletin of the Society of Pharmacological and Environmental Pathology*, **3**, 3–11.

TUCKER, M. J. (1975) Carcinogenic action of quinoxaline 1,4-dioxide in rats, *Journal of the National Cancer Institute*, **55**(1), 137–45.

8

The Haemopoietic and Lymphatic Systems

8.1 Bone Marrow

8.1.1 Non-neoplastic Changes

In the AP rat the cellularity of the bone marrow is examined in sections of sternum and femur as these bones show the most consistent cellularity in rats (Cline and Maronpot, 1985; Wright, 1989). Cytology is evaluated by examination of femoral marrow smears; these are taken from all animals at necropsy, except for any which died, where autolysis prevents adequate evaluation. The smears are not examined unless there are changes in the haematology, or the morphology of the haemopoieitic organs, which require further investigation.

Atrophy

The most common change in bone marrow in the AP rat is atrophy. The bone marrow becomes less cellular and there is replacement of the declining numbers of marrow cells by fat (Figure 63). This is more marked in females but, in 24 month studies, the incidences of significant fat replacement in males and females is 2.4 and 4.6 per cent, respectively. Between 24 and 36 months the incidence increases to 10 and 25 per cent. It is known that marrow cellularity does decline with age (Cline and Maronpot, 1985) and there is a sex difference in the effects of hormones on the marrow. The balance between the production of erythroblasts and lymphoblasts is dependant, in part, on the actions of erythropoietin, an erythropoiesis-stimulating factor, produced by the kidney (Stohlman *et al.*, 1968). Secretion of erythropoietin is stimulated by androgens (Gordon *et al.*, 1968), but in females oestrogens depress erythropoietin

123

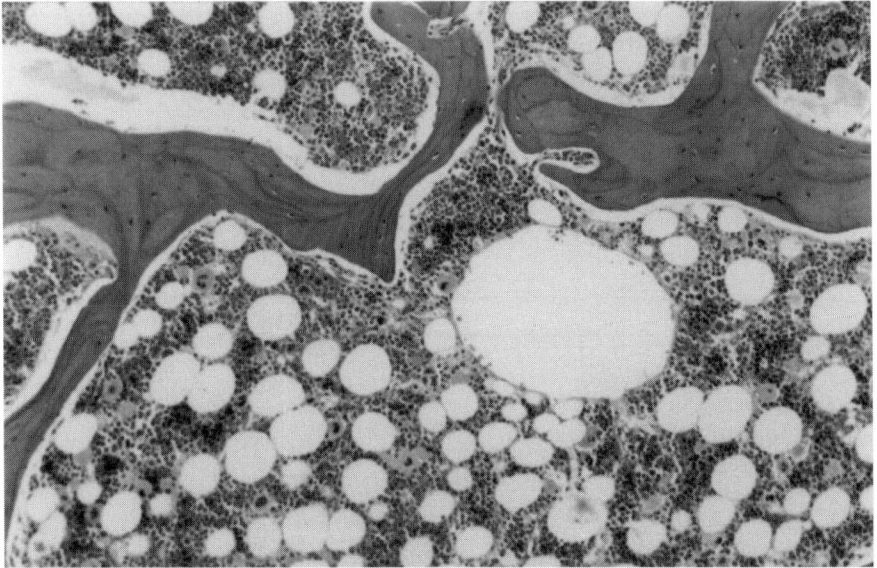

Figure 63 Fat replacement of marrow in the femur of a 24 month old AP female rat. ×32, H&E

production, and this is thought to account for the slightly lower red blood cell counts in females. The general effect is that androgens stimulate and oestrogens depress the formation of blood-forming elements.

Myelofibrosis and hyperplasia

Myelofibrosis is a rare condition seen in animals with hyperplasia of the parathyroid secondary to renal disease. Hyperplasia of different marrow cells is also rare, but it is seen as a result of a variety of inflammatory and neoplastic conditions. This low incidence of non-neoplastic marrow changes is common to all strains of rat (Stromberg, 1992).

8.1.2 Neoplastic Changes

Leukaemia in the AP rat is diagnosed only when the peripheral blood is involved. Both myeloid and lymphatic leukaemias have been observed, with a maximum overall incidence for all types of 2 per cent; myeloid is more common on a ratio of 3:1. Monocytic (large granular cell) leukaemia is less common and is discussed under spleen since this is the organ in which it originates, and in our experience the peripheral blood is only involved infrequently. The numbers of animals which have developed the other types of leukaemia are shown in Table 8.1. Myeloid leukaemias are distributed equally between the sexes but the lymphatic leukaemias were only found in males, and

Table 8.1 Incidence of leukaemia in the AP rat

Animal	Age in months	Leukaemia[a]	Year
Female	24	Myeloblastic	1961
Male	24	Myeloblastic	1969
Male	18	Lymphoblastic	1969
Male	12	Myeloblastic	1971
Female	23	Myeloblastic	1971
Male	26	Myeloblastic	1971
Female	26	Myeloblastic	1971
Male	18	Myeloblastic	1973
Male	22	Myeloblastic	1973
Female	13	Myeloblastic	1974
Male	14	Myeloblastic	1975
Male	21	Myeloblastic	1975
Female	26	Myeloblastic	1980
Male	17	Lymphoblastic	1981
Male	18	Lymphoblastic	1984
Male	14	Lymphoblastic	1984
Male	17	Lymphoblastic	1985
Female	22	Myeloblastic	1988
Female	24	Myeloblastic	1990

[a] Incidence of myeloid and lymphatic leukaemias in a database of 8880 control animals (including 2800 males and 2500 females in 2 year studies) used in toxicology studies between 1960 and 1994. Highest incidence in any study was 2%.

they tended to occur in young animals. The leukaemias were diagnosed from blood and marrow smears but, where blood was sampled, total white cell counts for lymphatic leukaemias were over 30 000 and for myeloblastic over 100 000. The clinical course of the leukaemias was rapid: from the first signs of weight loss it was usually only a few weeks to a moribund state. Only two females with myeloblastic leukaemia demonstrated the characteristic green colouration of tissues associated with chloroleukaemia. Lymphatic leukaemia was more widely disseminated in tissues. The cell type was a uniform large lymphoblast with no mature lymphocytes and usually a high mitotic rate; almost all organs were infiltrated, and the bone marrow completely replaced by leukaemic cells. In the liver, distribution had a distinct periportal appearance, and in the spleen the lymph follicles were replaced first, and then the leukaemia extended into the red pulp. Virtually all lymph nodes, and the thymus, were replaced by leukaemic cells and infiltration was also seen in two tissues which have never been infiltrated by the myeloid leukaemias: the choroid layer of the eye and the brain parenchyma.

The myeloblastic leukaemias appeared less aggressive and tissue infiltration was less extensive. In all cases the liver had a diffuse sinusoidal infiltration by

125

large myeloblasts, but varying proportions of more mature myeloid cells, including metamyelocytes and mature polymorphs, were present. They were never of a sufficiently high proportion to suggest a myelocytic myeloid leukaemia. The splenic pulp was also always infiltrated but the bone marrow usually showed areas of infiltration rather than the complete replacement seen in lymphatic leukaemias. These types of leukaemia have been induced in the AP rat by a potent leukaemogenic agent, β-chloroethylamine (Tucker, 1968; Leonard, 1968).

Other strains have also been reported to have a low incidence of these types of leukaemias (Swaen and Van Heerde, 1973) including the F344 (Stromberg, 1990), SD rats (Frith, 1988), and other Wistar strains (Schreiner and Will, 1962).

8.2 *Spleen*

8.2.1 *Non-neoplastic Changes*

Spleen weights

The weight of the spleen at different time points is shown in Table 8.2. Unlike most other organs the absolute weight of the spleen increases up to 58 weeks while the relative weight decreases over the same time period. Weights have not been recorded in 2 year studies as they are frequently affected by neoplastic and other diseases. The spleen weights appear to be in agreement with those described by Losco (1994) for other outbred SD and Wistar strains.

Accessory spleens

Small accessory spleens have been found, rarely, in the tail of the pancreas of the AP rat. They have been observed in similar low incidence in other strains, but should be distinguished from splenic foci from ruptured spleens, which may be found anywhere in the abdominal cavity (Wolf and Neiman, 1989).

Table 8.2 Spleen weights in the AP rat

| Age (weeks) | Number/ sex | Mean weight of spleen | | | |
| | | Weight (g) | | Relative weight (% body weight) | |
		Male	Female	Male	Female
12	10	0.88	0.64	0.25	0.27
34	20	1.11	0.81	0.19	0.25
58	25	1.31	0.96	0.16	0.21

Extramedullary haematopoiesis

The spleen of the rat responds rapidly to increased demand for haematopoietic cells. This was studied in hypoxic rats where an immediate erythropoiesis occurred on development of hypoxia; it reached a maximum in 2 to 4 weeks, and returned to normal within 4 weeks of the removal of the hypoxia (Stutte *et al.*, 1986). Extramedullary haematopoiesis is a normal feature of the rat spleen in young animals and declines with age. It can increase significantly in inflammatory or neoplastic diseases and increase splenic weight two to five fold (Ward, 1990a).

Congestion/macrophages/osseous metaplasia

Congestion is also a common observation in the spleen. It can be an agonal change related to the method of anaesthesia rather than to a frank venous congestion from raised venous pressure. The spleen has a large population of macrophages in the pulp and these are concerned with the removal of damaged or senile red cells. Thus haemosiderin pigment is a common finding in sinusoidal macrophages (Ward and Reznik-Schuller, 1980) and increases with age as conditions develop where red cell destruction is increased. Osseous metaplasia has been seen in a few AP rats with marked atrophy of the spleen.

Cysts and siderofibrosis

The capsule of the spleen can develop one or more serous cysts which may be quite large, but appear to have no clinical significance. Siderofibrosis of the capsule, which is a focal, or occasionally diffuse, condition of fibrous thickening with large numbers of pigment laden macrophages, is an uncommon condition thought to occur as a response to peritoneal inflammation or tumours.

Lymphoid hyperplasia/depletion

The histological appearance of the spleen varies markedly in older animals as the white pulp of the spleen is increased or depleted as a response to various diseases. In older rats the T cell zone around the follicular arteries (the periarteriolar lymphoid sheath – PALS) becomes depleted of cells (Cheung *et al.*, 1981) but this can also occur in viral infections. Hyperplasia of the B cell zone which adjoins the PALS is seen in young rats exposed to a new antigen. Secondary germinal centres are formed and there is plasmacytosis. Hyperplasia or enlargement of the marginal zones (which are the pale staining areas at the periphery) are most likely to be early neoplastic infiltration. Necrosis and atrophy occur rarely, consequent to haemorrhagic shock, and are thought to result from corticosteroid release (Gopinath *et al.*, 1987).

Arteritis

Arteritis of the follicular vessels is one of the more common sites for the disease in the AP rat, but infarcts have only been observed in two animals. The incidence of splenic arteritis is less than 10 per cent in any study and is more common in males. The disease is seen more frequently in the spleen of SD rats (Anver *et al.*, 1982) where it is found in over 15 per cent of rats over 24 months of age.

8.2.2   Neoplastic Changes

Mesenchymal tumours

Neoplasms of the spleen are very rare and have not exceeded 1 per cent in any single study. The overall incidence is higher in males, but with such small numbers in total it is not possible to say if there is a real difference between the sexes. The types and incidence of tumours in the spleen are shown in Table 8.3. The histological appearance of the mesenchymal tumours is similar to that described in other organs. The lymphoblastic tumours are more common as secondary infiltrations from tumours in lymph nodes but a few have been observed as primary tumours in the spleen. They were both nodular tumours rather than the follicular infiltrating secondary tumours, and were composed of uniform large lymphoblasts showing frequent mitoses; neither had spread to other organs.

Table 8.3   Incidence of tumours of the spleen in the AP rat

Type of tumour	Incidence[a]	Year observed
Fibroma	Male aged 31 months	1962
	Male aged 31 months	1962
Fibrosarcoma	Male aged 25 months	1973
	Male aged 24 months	1974
	Female aged 26 months	1975
Angioma	Male aged 26 months	1962
Angiosarcoma	Male aged 24 months	1962
	Female aged 24 months	1986
Liposarcoma	Male aged 26 months	1982
Lymphoblastic lymphoma	Female aged 26 months	1974
	Female aged 26 months	1984
Monocytic leukaemia	Male aged 26 months	1982
	Male aged 22 months	1990

[a] Incidence of tumours from a database of 8880 control animals (including 2800 males and 2500 females in 2 year studies) used in toxicology studies between 1960 and 1994.

Monocytic leukaemia

Monocytic leukaemia follows the pattern of the disease described for other strains, where the leukaemia is very common. The spleen and liver were enlarged and both were extensively infiltrated by large granular lymphocytes. In the spleen the leukaemia infiltrated the red pulp and was accompanied by depletion of the lymph follicles, while the liver showed a diffuse sinusoidal infiltration and nodular hyperplasia of the liver parenchyma. Bone marrow and peripheral blood were not involved. The incidence of this leukaemia is very low compared with the F344 rat, where levels up to 50 per cent have been recorded (Goodman *et al.*, 1979) and where it is a major cause of death. In other strains lower incidences of monocytic leukaemias have been recorded, but it is still a more common tumour than in the AP rat. All other types are uncommon, as in the AP rat (Moloney *et al.*, 1969; Abbott *et al.*, 1983) and other Wistar strains (Kroes *et al.*, 1981).

8.3 Lymph Nodes

8.3.1 *Non-neoplastic Changes*

The mesenteric and submandibular lymph nodes are examined routinely in toxicology studies in the AP rat, but any lymph node with macroscopic abnormality is also taken for histological examination. The exception to this rule is animals with generalised lymph node enlargement due to neoplasia, in which case only selected nodes are examined.

Hyperplasia

In a typical immune response in the lymph node the germinal centres of the cortex become enlarged and prominent (Figure 64). The germinal centres are normally populated by B lymphocytes, but in immune stimulation secondary follicles may be formed which are populated by large pale staining lympho-blasts and macrophages (Figure 65). The stimulus for this hyperplasia is thought to be initiated by paracortical macrophages ingesting an antigen; they then migrate to the cortex and stimulate antigen-sensitive lymphocytes to proliferate. It can occur as a reaction to infection, but over the last 40 years there has been an increase in reactive hyperplasia of the lymph nodes, presumably because of increasing exposure to new, unfamiliar antigens. The incidence of reactive hyperplasia in various lymph nodes in 600 control animals in a 2 year study is shown in Table 8.4. Hyperplasia is much more frequent in the submandibular node than in the mesenteric node, and this is a consistent feature of all toxicology studies in the AP rat. Follicular hyperplasia in the submandibular lymph node occurs in young rats, but in older animals there is also plasmacytosis of the paracortical and medullary areas. This has also been recorded in SD rats by Losco and Harleman (1994).

129

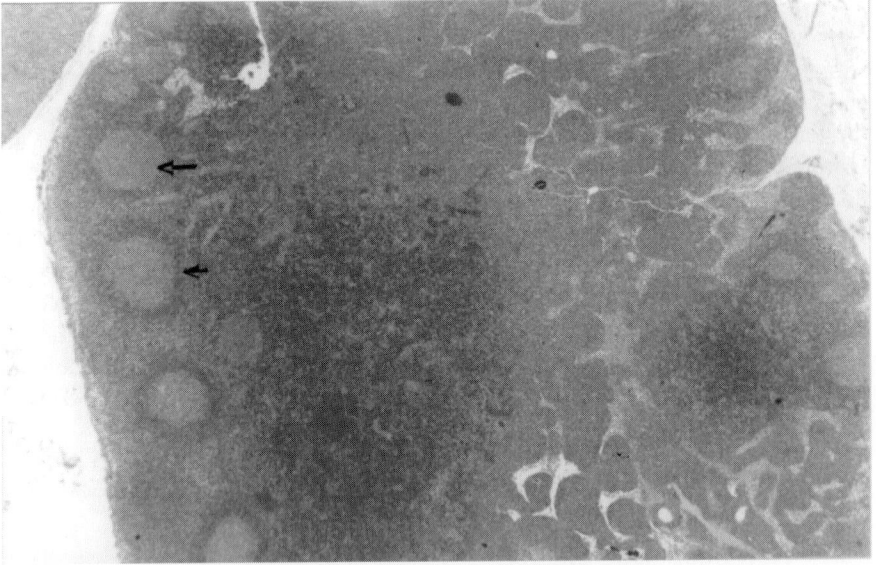

Figure 64 Reactive lymphoid hyperplasia of a submandibular lymph node in a male AP rat, showing enlarged and prominent germinal centres. ×8, H&E

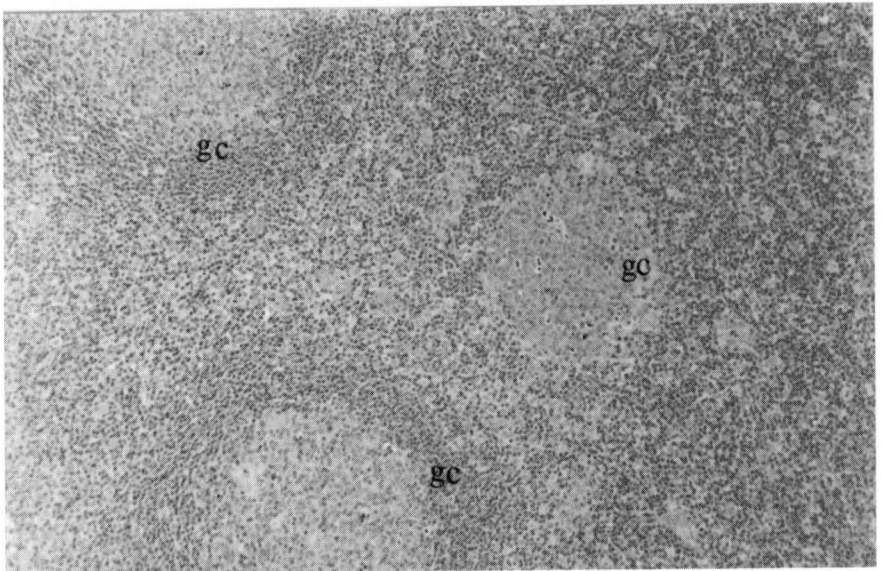

Figure 65 Reactive lymphoid hyperplasia with enlarged germinal centres (gc) populated by lymphoblasts and macrophages. ×80, H&E

Table 8.4 Incidence of lymph node hyperplasia in the AP rat

Lymph node[a]	Incidence[b] of reactive lymphoid hyperplasia	
	Male	Female
Mesenteric	5/300 (1.6%)	2/300 (0.6%)
Submandibular	102/300 (34%)	123/300 (41%)
Axillary	3/19	2/10
Cervical	4/11	6/14
Bronchial	0/1	NE
Mediastinal	0/45	1/36
Thoracic aortic	0/1	NE
Cisternal	0/22	1/3
Superficial inguinal node	0/5	2/2
Deep inguinal node	7/41	9/20
Iliac	0/3	2/2
Lumbar aortic node	NE	2/2
Pancreatic	5/27	4/19
Popliteal	2/5	2/2
Renal	2/118	0/62

[a] Mesenteric and submandibular nodes examined in all animals; all other lymph nodes were only examined when macroscopically abnormal.
[b] Incidence in a 2 year study of 600 control AP rats.
NE = None examined.

A variety of other conditions occur in lymph nodes, some of which may be regarded as normal variations in the function of nodes. Reporting of such changes, particularly when they are minimal, varies markedly among pathologists. The incidence of some of these changes in the AP rat is shown in Table 8.5.

Inflammation

Lymphadenitis is a rare (<1 per cent) condition in the AP rat. The nodes involved are enlarged by oedema and inflammatory cell infiltrates, and in chronic conditions there is fibrosis. The inflammatory cell infiltrates are usually a mixture of neutrophils and lymphocytes with varying proportions of macrophages. The normal architecture of the node is not disturbed, in contrast to neoplastic infiltrates. The nodes most frequently involved are those which drain the skin, and this has also been observed by Ward (1990b).

Pigmentation and haemorrhage

After enlargement due to reactive lymphoid hyperplasia, the most common macroscopic observation in the lymph node is discoloration. This is due to

Table 8.5 Incidence of various changes in the lymph nodes of AP rats

Change	Lymph node[a]	Number of animals with change[b] Male	Female
Lymphadenitis	Submandibular	0/300	0/300
	Mesenteric	1/300	1/300
	Axillary	1/19	0/10
	Cervical	1/11	0/14
	Deep inguinal	2/41	0/20
	Pancreatic	1/27	0/19
Haemosiderosis/	Submandibular	19/300	19/300
erythrocytosis	Mesenteric	77/300	57/300
	Axillary	8/19	3/10
	Cervical	6/11	5/14
	Cisternal	6/22	1/3
	Deep inguinal	10/45	4/20
	Mediastinal	28/45	30/36
	Pancreatic	17/27	9/19
	Renal	78/118	37/62
Cystic degeneration	Submandibular	131/300	89/300
	Mesenteric	0/300	0/300
	Axillary	5/19	3/10
	Cervical	6/11	5/14
	Cisternal	14/22	1/3
	Superficial inguinal	2/5	0/2
	Deep inguinal	16/41	5/20
	Iliac	1/3	0/2
	Mediastinal	14/45	3/36
	Pancreatic	2/27	3/19
	Popliteal	3/5	1/2
	Renal	36/118	2/62
Histiocytosis	Submandibular	1/300	0/300
	Mesenteric	7/300	21/300
	Axillary	2/19	1/10
	Cisternal	1/22	0/3
	Deep inguinal	2/41	0/20
	Pancreatic	2/27	2/19
	Renal	0/118	1/62

Table 8.5 *(continued)*

Change	Lymph node[a]	Number of animals with change[b]	
		Male	Female
Tumour metastases[c]	Cervical	0/11	2/14
	Cisternal	1/22	0/3
	Superficial inguinal	1/5	0/2
	Deep inguinal	2/41	1/20
	Mediastinal	2/45	2/36
	Pancreatic	0/27	1/19
	Renal node	2/118	2/62
Arteritis	Bronchial	1/1	NE
Mineralisation	Submandibular	1/300	0/300
	Mesenteric	1/300	0/300
Infarct	Submandibular	1/300	0/300

[a] Submandibular and mesenteric nodes were examined in all animals, other nodes were examined when macroscopically abnormal.

[b] Incidence in a 2 year study of 600 untreated AP rats.

[c] Metastatic tumours include metastases from leukaemias in cervical, renal (3), mediastinal, pancreatic, and superficial and deep inguinal nodes; thyroid tumour in cervical, skin tumour in cisternal and mediastinal, seminal vesicle tumour in deep inguinal, spleen tumour in mediastinal nodes.

NE = None examined.

haemorrhage (sinus erythrocytosis) or to haemosiderin deposits. In the AP rat the nodes most frequently affected by these changes are the renal and mediastinal nodes.

Cystic degeneration

Cystic degeneration (cystic lymphangiectasis) occurs in up to 10 per cent of AP rats in 2 year studies. It is a lesion of old rats and does not occur in animals less than 18 months of age. The cystic area may be a large single focus (Figure 66) containing pale pink fluid; or it may be a multifocal lesion of small cysts. Burek (1978a) reported a 15 per cent incidence of cystic degeneration in 35 month old (WAGxBN) F1 rats, and Losco and Harleman (1994) recorded a 5–26 per cent incidence in F344 and SD rats. They recorded the mediastinal and mesenteric nodes as the most common site whereas in the AP rat it is the submandibular node.

Histiocytosis

Histiocytosis with granuloma formation (Figure 67) is a non-specific change

133

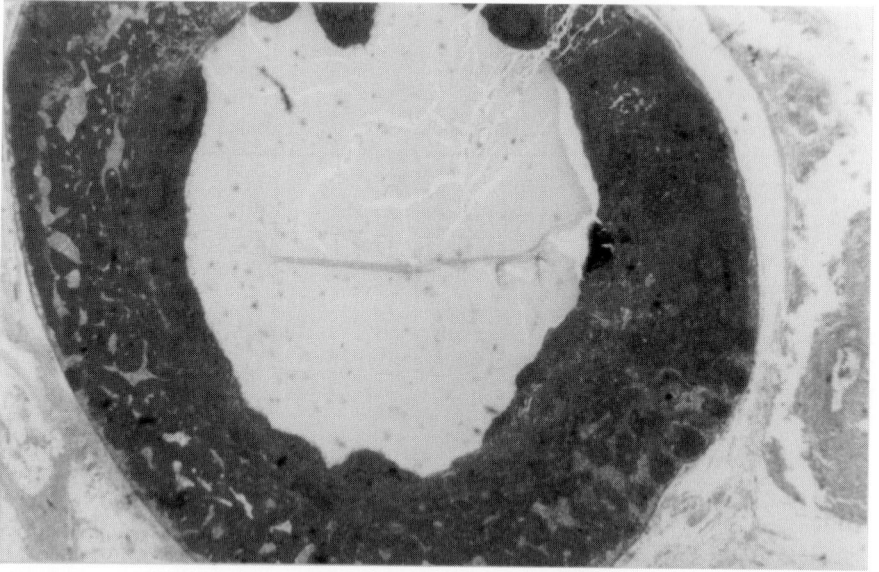

Figure 66 Cystic degeneration in a submandibular lymph node of a male AP rat. ×8, H&E

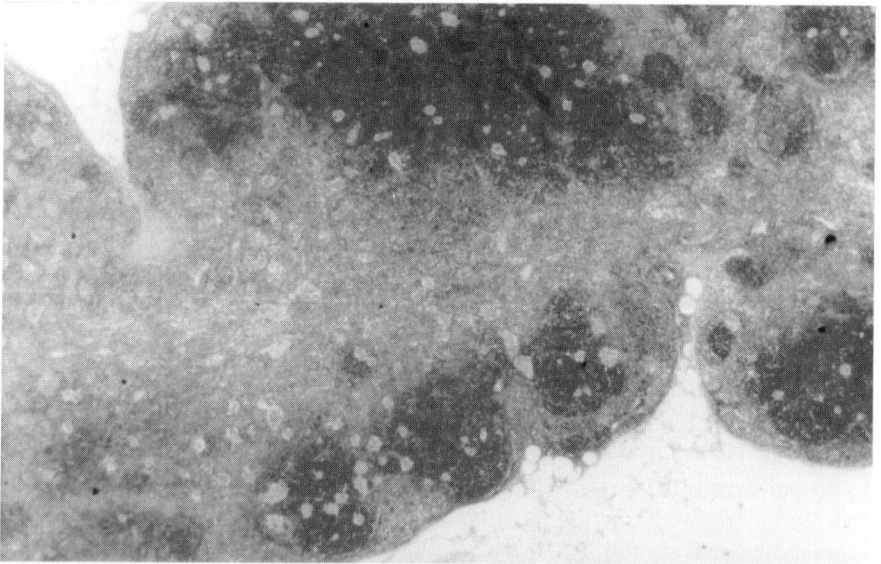

Figure 67 Histiocytosis of the mesenteric lymph node in a male AP rat. Multifocal aggregates of pale macrophages can be seen, chiefly in paracortical areas. ×8, H&E

seen most commonly in the mesenteric node of the AP rat. There are multifocal aggregates of macrophages in the paracortical and medullary areas. This change is rare in rats less than 12 months old.

8.3.2 Neoplastic changes

The most common neoplasm in the lymph node of the AP rat is a benign angioma (Figure 68) of the mesenteric node. These angiomas have irregular sized blood-filled spaces lined by plump endothelial cells, a variable amount of collagenous stroma and little mitotic activity. The incidence of this tumour is shown in Table 8.6 and has varied markedly over the years, but has always been more common in males. This high incidence is a feature of Wistar strains (Deerberg *et al.*, 1982) and unlike the low incidences recorded for the SD and F344 rats (Goodman *et al.*, 1979; Frith, 1988). The aetiology of the tumour is not known but Rehm *et al.* (1984) noted that the incidence was much lower in breeding animals than in virgin animals. This might indicate that some hormonal factor is involved. A few angiosarcomas of the mesenteric lymph node have been observed. These tumours still show some blood-filled spaces but the lining is very irregular and composed of pleomorphic cells, often in solid masses, with numerous mitotic figures. Local invasion outside the node has been found, but no metastases, and the maximum incidence observed has been less than 1 per cent in any study.

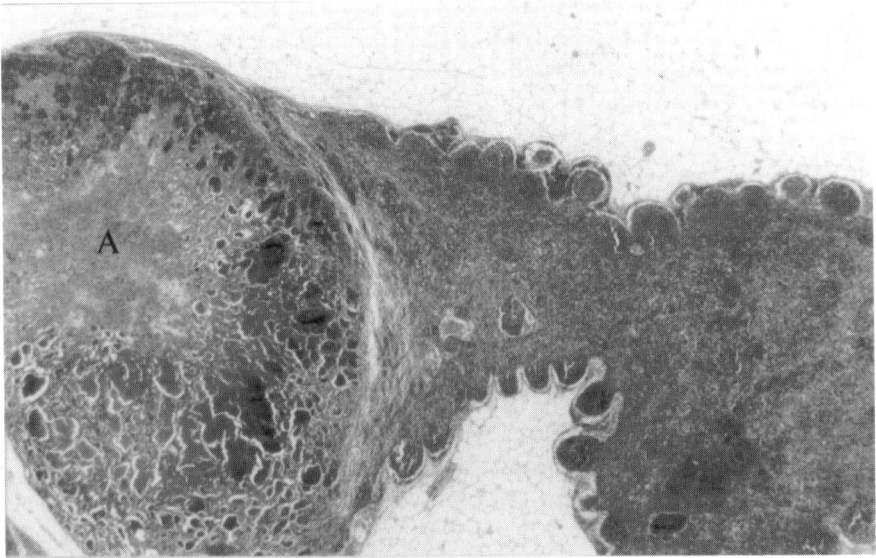

Figure 68   Angioma (A) in the mesenteric lymph node of a 26 month old male AP rat. ×8, H&E

135

Table 8.6 The incidence of mesenteric lymph node angiomas in the AP rat

Year[a]	Incidence of angiomas of mesenteric lymph node (%)	
	Males	Females
1963	4	0
	4	0
1965	10	0
1967	4	4
1971	0	2
1973	10	10
1974	7	0
1975	5	0
1976	2	0
	14	0
1980	7	1
1981	13	3
1983	10	3
1984	15	1
1986	9	2
1987	3	0
1990	8	2
1992	12	2

[a] Year carcinogenicity study was completed. Six studies had a zero incidence.

The histological types of other primary lymphoid tumours of the lymph node are shown in Table 8.7 with the maximum incidence observed for each type. The incidence of all types is low, as it is in most strains (Coleman *et al.*, 1977; Goodman *et al.*, 1979, 1980; Kroes *et al.*, 1981; Frith, 1988). The malignant lymphomas include several cell types. Lymphoblastic lymphomas are composed of uniform, large lymphoblastic cells showing a high mitotic rate. It is usually disseminated widely in the tissues, particularly in other nodes, the thymus, spleen and liver. Bone marrow and peripheral blood were never involved. These tumours are said to be of B or T cell origin (Pattengale and Frith, 1986). It was the most common type of lymphoma in aged SD rats (Frith, 1988). The only lymphocytic lymphoma observed was in the mesenteric lymph node of a female aged 26 months. The node was greatly enlarged and composed of small, mature cells; mitoses were infrequent and the tumour had not spread to other organs except the adjacent adipose tissue. The histiocytic lymphomas constitute the most common type of lymphoma overall, but care should be taken to distinguish them from metastases of soft tissue

Table 8.7 Incidence of lymph node tumours (excluding angiomas) in the AP rat

	Highest incidence observed[a] (%)	
Type of tumour	Male	Female
Lymphoblastic lymphoma	1.0	0.7
Lymphocytic lymphosarcoma	0.5	0
Histiocytic lymphoma	2.0	2.0
Plasma cell lymphoma	0.3	0
Hodgkin's lymphoma	1.3	0

[a] Highest incidence observed in 24 carcinogenicity studies in the AP rat.

fibrohistiocytic sarcomas. The histiocytic sarcoma is a tumour of pleomorphic cells, some with abundant cytoplasm and some with clearly indented nuclei. The tumour infiltrates other nodes and liver, spleen and lungs. Three plasma cell lymphomas have been found in abdominal nodes. The tumours were of mature plasma cells with basophilic cytoplasm and round nuclei with the characteristic marginated chromatin. Two were found in females aged 5 and 19 months and one in a male aged 26 months. The one tumour designated as a Hodgkin's-like lymphoma had infiltrated several nodes, including the cervical and mesenteric nodes. It was a mixed cell tumour with a high proportion of multinucleate giant cells and large cells with indented nuclei.

8.4 Thymus

8.4.1 *Non-neoplastic Changes*

Congenital abnormalities

Ectopic thymus is found occasionally in the thyroid gland, and cysts lined by a squamous or ciliated epithelium are frequent findings.

Thymic weights

Weights of the thymus in 3 and 8 month old AP rats (i.e. at the end of 1 and 6 month studies) are shown in Table 8.8. The thymus is not weighed in studies of longer duration as it cannot be identified consistently and weighed after 6 months. In males the thymus reduces by approximately 40 per cent between 3 and 8 months of age, but in females the reduction is only 9 per cent. This has been seen in other Wistar rats (Kuper *et al.*, 1986).

137

Table 8.8 Thymus weights in the AP rat

Age in months[a]	Number/ sex	Mean weight of thymus (g)	
		Male	Female
3	10	0.567	0.438
8	20	0.338	0.399

[a] The thymus gland is not weighed in studies longer than 6 months due to individual variation in the rate of involution.

Atrophy

Atrophy of the thymus is a distinctly age-related change although it may undergo very rapid involution if the animal is subjected to stress, such as a virus infection. In the young AP rat, up to 6 months of age, the thymus has an extensive cortex populated with small lymphocytes, which contrasts with the paler staining medulla which has fewer lymphocytes (Figure 69). As the animal ages, the lymphocytes are depleted and the cortical area, in particular, becomes much smaller by 12 months (Figure 70). In the AP rat the thymus is often not identified at necropsy in animals killed at the end of a 2 year study, and

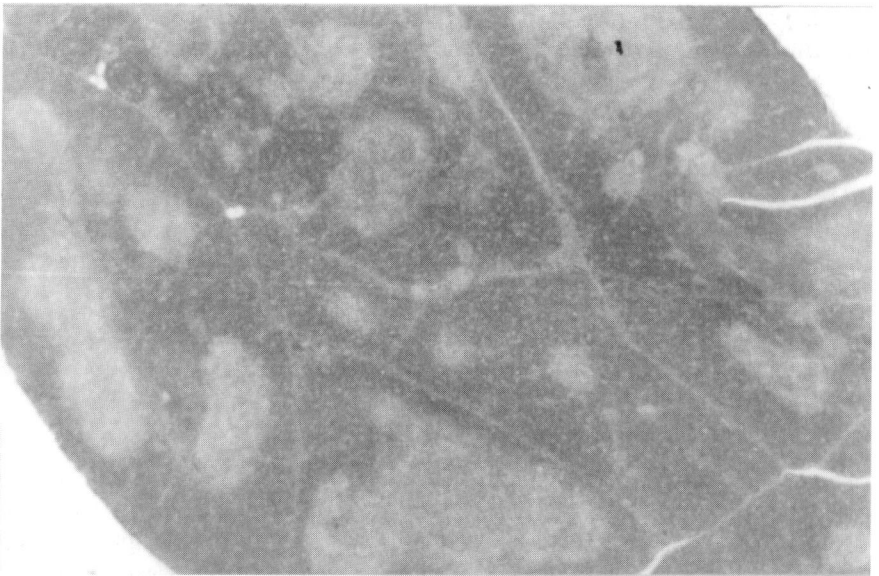

Figure 69 Thymus of a 3 month old female AP rat: the gland has an extensive darkly stained cortex populated by lymphocytes and a paler staining medullary area. ×8, H&E

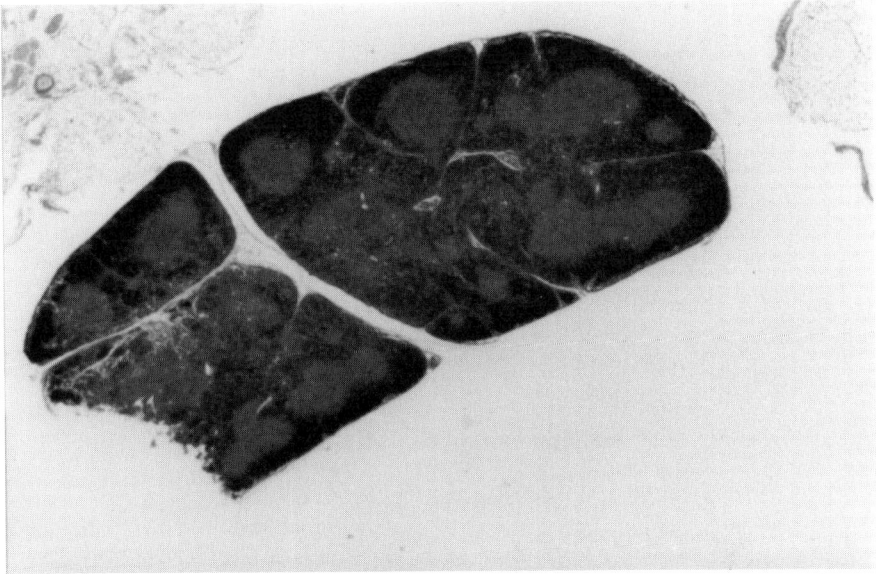

Figure 70 Thymus of a 14 month old AP rat showing moderate atrophy with a marked decrease in the size of the cortex. ×8, H&E

microscopic examination shows only small irregular strands of tissue with no distinction of cortex and medulla. The rate and extent of thymic involution is strain, age and sex dependant (Kuper *et al.*, 1986). They observed that there was a correlation between thymic involution and ovarian atrophy, indicating the relationship between gonadal steroids and the thymus.

Necrosis

Lymphocytolysis has been seen in AP rats with virus infections such as Sendai and rat corona virus. Necrosis is primarily confined to cortical lymphocytes, while the medulla was little affected.

8.4.2 Neoplastic Changes

Benign thymoma

The most common tumour in the thymus of the AP rat is a benign lymphocytic thymoma. This tumour is a mixture of epithelial cells and mature lymphocytes. The normal architecture of the thymus is lost, although there are pale staining medulla-like areas between the darker staining strands of lymphocytes (Figure 71) and fibrous trabeculae traverse the tumour dividing it into pseudo lobules. In some tumours the epithelial components are more prominent and there are fewer

139

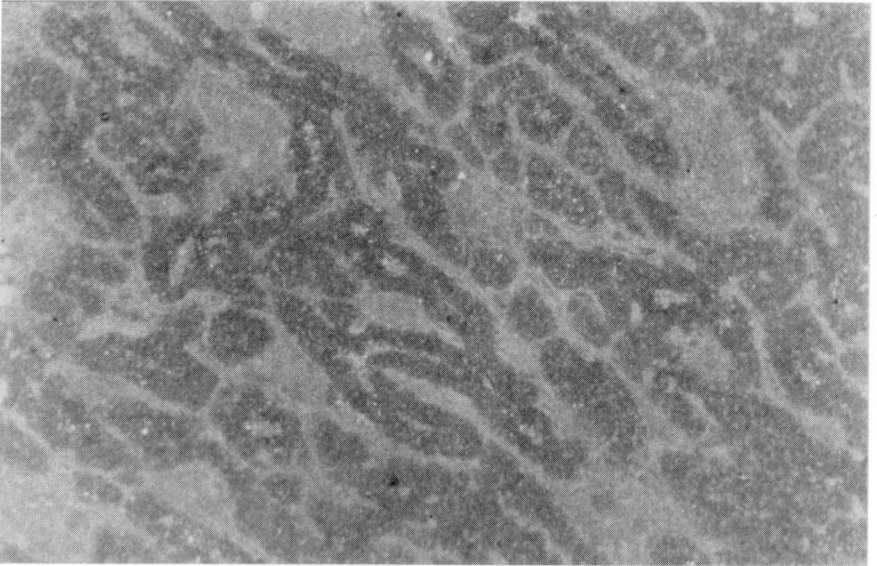

Figure 71 Benign thymoma in a 24 month old AP rat showing loss of normal architecture and pale staining medullary-like areas and strands of lymphocytes. ×80, H&E

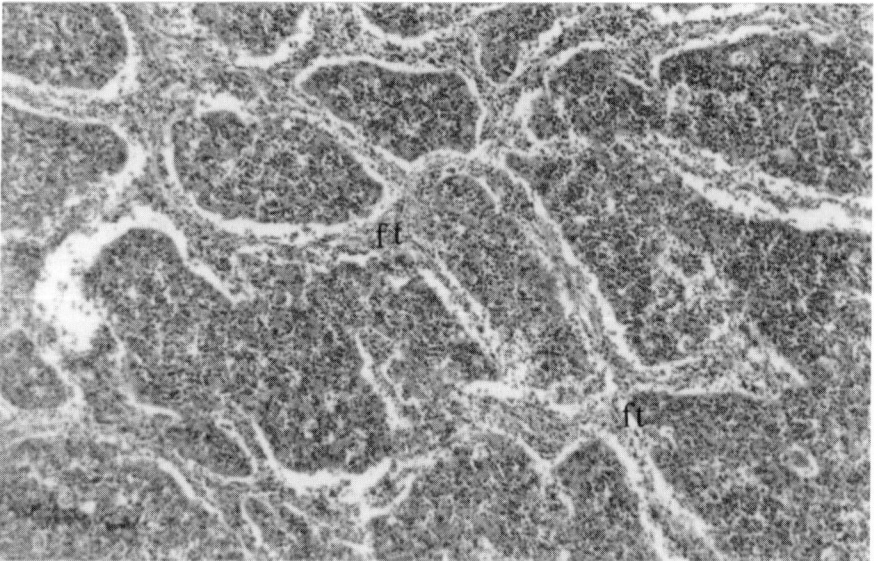

Figure 72 Benign thymoma in a 24 month old female AP rat, showing fibrous trabeculae (ft). ×80, H&E

Table 8.9 Incidence of benign lymphocytic thymomas in the AP rat

Year	Percentage incidence of thymomas[a]	
	Male	Female
1963	0.5	1
1969	12	12
1971	4	0
1973	5	10
1974	2	6
1975	2	9
	1	8
1976	0	4
1980	1	10
1984	2	7
1986	0	12
1988	2	8
1992	3	6

[a] Incidence of benign thymomas in 24 carcinogenicity studies in the AP rat. Studies with a zero incidence in both sexes are excluded.

lymphocytes (Figure 72). These thymomas do not metastasise and are not locally invasive as many are encapsulated. They may become very large, and respiratory distress is a clinical sign which precedes death caused by compression of thoracic organs. In the AP rat the incidence of this tumour is very variable from one study to another, but shows no general trend to an increase in number over the last 40

Table 8.10 Incidence of thymic tumours (excluding thymomas) in the AP rat

Type of thymic tumour	Number observed[a]	
	Male	Female
Lymphoblastic lymphosarcoma	3	1
Adenoma	4	2
Cystadenoma	1	0
Squamous carcinoma	3	0

[a] Number observed in a database of 8880 control AP rats (including 2800 males and 2500 females in 2 year studies) used in toxicology studies between 1960 and 1994.

141

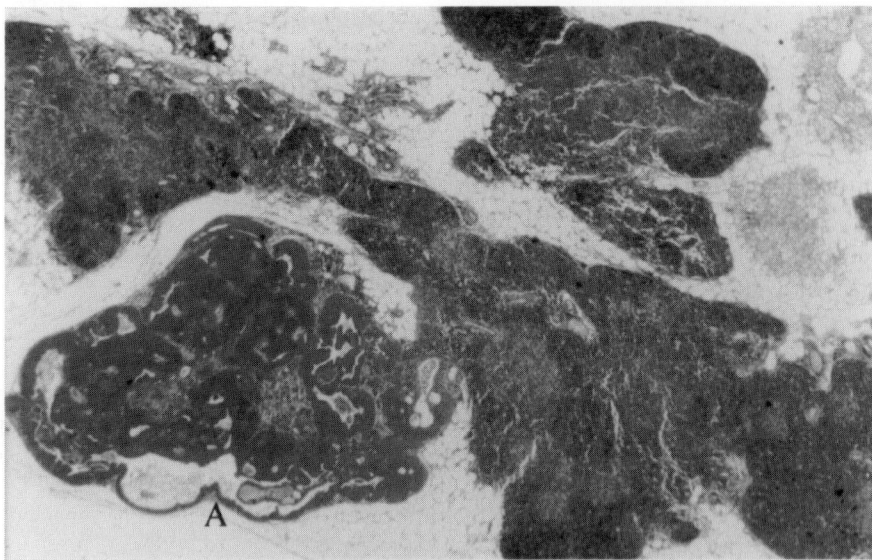

Figure 73   Adenoma (A) of the thymus in a male AP rat. ×8, H&E

years (Table 8.9), but they are more frequent in females. The incidence of these tumours in other strains appears to be much lower (Burek, 1978b; Altman, 1985; Kuper *et al.*, 1986; Naylor *et al.*, 1988).

Other types of thymic tumour observed in the AP rat are shown in Table 8.10. They occur much less frequently than thymomas; the maximum incidence of all types in any study is less than 2 per cent. The lymphosarcomas were only designated as primary tumours of the thymus when there was no major involvement of lymph node or spleen. They were all lymphoblastic in cell type and showed local extension into mediastinal adipose tissue. The adenomas were all well-differentiated small tumours of duct-like or small cystic structure with variable amounts of collagenous stroma (Figure 73). The squamous carcinomas had the histological appearance of the tumour in the skin. They are thought to have arisen from small squamous cysts which are not infrequent in the thymus.

8.5   References

ABBOTT, D. P., PRENTICE, D. E. and CHERRY, C. P. (1983) Mononuclear cell leukaemia in aged Sprague-Dawley rats, *Veterinary Pathology*, **20**, 434–9.

ALTMAN, Ph. L. (1985) in *Pathology of Laboratory Mice and Rats*, pp. 295–8, Virginia: Pergamon.

ANVER, M. R., COHEN, B. J., LATTUADA, C. P. and FOSTER, S. J. (1982) Age associated lesions in barrier-reared male Sprague-Dawley rats: a comparison between Hap:(SD) and Crl:COBS®CD®(SD) stocks, *Experimental Aging Research*, **8**, 3–24.

BUREK, J. D. (1978a), *Pathology of Aging Rats*, pp. 111–12, Palm Beach, Florida: CRC Press.

BUREK, J. D. (1978b), *Pathology of Aging Rats*, pp. 113, Palm Beach, Florida: CRC Press.

CHEUNG, H. T., VOVOLKA, J. and TERRY, D. S. (1981) Age and maturation-dependant changes in the immune system of Fischer 344 rats, *Journal of the Reticuloendothelial Society*, **30**, 563–72.

CLINE, J. M. and MARONPOT, R. R. (1985) Variations in the histologic distribution of rat bone marrow cells with respect to age and anatomic site, *Toxicologic Pathology*, **13**, 349–55.

COLEMAN, G. L., BARTHOLD, S. W., OSBALDISTON, G. W., FOSTER, S. J. and JONES, A. M. (1977) Pathologic changes during aging in barrier-reared Fischer 344 male rats, *Journal of Gerontology*, **32**, 258–78.

DEERBERG, F., RAPP, K. G. and REHM, S. (1982) Mortality and pathology of Han:Wist rats depending on age and genetics, in GAERTNER, K. H. and STOLTE, H. (Eds), *Experimental Biology and Medicine: Monographs on Interdisciplinary Topics*, Vol. 7, pp. 63–71, Basel and New York: Karger.

FRITH, C. H. (1988) Morphological classification and incidence of haemopoietic lesions in the Sprague-Dawley rat, *Toxicologic Pathology*, **16**, 451–7.

GOODMAN, D. G., WARD, J. M., SQUIRE, R. A., CHU, K. C. and LINHART, M. S. (1979) Neoplastic and non-neoplastic lesions in aging F344 rats, *Toxicology and Applied Pharmacology*, **48**, 237–48.

GOODMAN, D. G., WARD, J. M., SQUIRE, R. A., PAXTON, M. B., REICHARDT, W. D., CHU, K. C. and LINHART, M. S. (1980) Neoplastic and non-neoplastic lesions in aging Osborne-Mendel rats, *Toxicology and Applied Pharmacology*, **55**, 433–47.

GOPINATH, R., PRENTICE, D. E. and LEWIS, D. J. (1987), in GRESHAM, G. A. (Ed.), *Atlas of Experimental Toxicological Pathology*, pp. 130–1, Boston: MTP Press.

GORDON, A. S., MIRAND, E. A., WENIG, J., KATZ, R. and ZANJANI, E. D. (1968) Androgen actions on erythropoiesis, *Annals of the New York Academy of Sciences*, **149**, 318–35.

KROES, R., GARBIS-BERKVENS, J. M., DE VRIES, Y. and VAN NESSELROOY, J. H. (1981) Histopathological profile of a Wistar rat stock including a survey of the literature, *Journal of Gerontology*, **36**, 259–79.

KUPER, C. F., BEEMS, R. B. and HOLLANDERS, V. M. H. (1986) Spontaneous pathology of the thymus in the aging Wistar (cpb.WU) rats, *Veterinary Pathology*, **23**, 270–7.

LEONARD, B. J. (1968) The leukaeogenic properties of β-chloroethylamine – ICI 42,464, *Proceedings of the European Society for the Study of Drug Toxicity*, X, pp. 183–90, ICS 181, Amsterdam: Excerpta Medica.

LOSCO, P. (1994) Normal development, growth and aging of the spleen, in MOHR, U., DUNGWORTH, D. L. and CAPEN, C. C. (Eds), *Pathobiology of the Aging Rat*, pp. 75–6, Washington: ILSI Press.

LOSCO, P. and HARLEMAN, H. (1994) Normal development, growth and aging of the lymph node, in MOHR, U., DUNGWORTH, D. L. and CAPEN, C. C. (Eds), *Pathobiology of the Aging Rat*, Vol. 1, pp. 49–74, Washington: ILSI Press.

MOLONEY, W. C., BOSCHETTI, A. E. and KING, V. (1969) Observations on leukaemia in Wistar Furth rats, *Cancer Research*, **29**, 938–46.

NAYLOR, P. H., KRINKE, G. J. and RUEFENACHT, H. J. (1988) Primary tumours of the thymus in the rat, *Journal of Comparative Pathology*, **99**, 187–203.

PATTENGALE, P. K. and FRITH, C. H. (1986) Contributions of recent research to the classification of spontaneous lymphoid cell neoplasms in mice, *CRC Critical Reviews in Toxicology*, **16**, 185–212.

REHM, S., DEERBERG, F. and RAPP, K. G. (1984) A comparison of life span and spontaneous tumor incidence of male and female Han:Wist virgin and retired breeders, *Laboratory Animal Science*, **34**, 458–64.

SCHREINER, A. W. and WILL, J. J. (1962) A transplantable spontaneous chloroleukaemia in the Wistar rat, *Cancer Research*, **22**, 757–60.

STOHLMAN, J. R., EBBE, S., MORSE, B., HOWARD, D. and DONOVAN, J. (1968) Regulation of erythropoiesis. XX: Kinetics of red cell production, *Annals of the New York Academy of Sciences*, **149**, 156–72.

STROMBERG, P. C. (1990) Haemopoietic neoplasms of Fischer 344 rats, in STINSON, S. F., SCHULLER, H. M. and REZNIK, G. (Eds), *Atlas of Tumor Pathology of the Fischer Rat*, pp. 505–26, Boca Raton, Florida: CRC Press.

STROMBERG, P. C. (1992) Changes in the hematologic system, in MOHR, U., DUNGWORTH, D. L. and CAPEN, C. C. (Eds), *Pathobiology of the Aging Rat*, pp. 15–24, Washington: ILSI Press.

STUTTE, H. J., SAKUMA, T., FALK, S. and SCHNEIDER, M. (1986) Splenic erythropoiesis in rats under hypoxic conditions, *Virchows Archiv [A]*, **409**, 251–61.

SWAEN, G. J. V. and VAN HEERDE, P. (1973) Tumours of the haematopoietic system, in TURUSOV, V. S. (Ed.), *Pathology of Tumours in Laboratory Animals*, Vol. 1, *Tumours of the Rat*, pp. 185–201, Lyon: IARC.

TUCKER, M. J. (1968) Observations relating to the carcinogenic action of pronethalol ('Alderlin'), *Proceedings of the European Society for the Study of Drug Toxicity*, Vol. X, 175–82.

WARD, J. M. (1990a) Classification of reactive lesions, spleen, in JONES T. C., WARD, J. M., MOHR, U. and HUNT, R. D. (Eds), *Haemopoietic System*, pp. 220–225, Berlin, Heidelberg and New York: Springer-Verlag.

WARD, J. M. (1990b) Classification of reactive lesions of lymph nodes, in JONES, T. C., WARD, J. M., MOHR, U. and HUNT, R. D. (Eds), *Haemopoietic System*, pp. 155–61, Berlin and Heidelberg: Springer-Verlag.

WARD, J. M. and REZNIK-SCHULLER, H. (1980) Morphological and histochemical characteristics of pigments in aging F344 rats, *Veterinary Pathology*, **17**, 678–85.

WRIGHT, J. A. (1989) A comparison of rat femoral, sternebral and lumbar vertebral bone marrow fat content by subjective assessment and image analysis of histological sections, *Journal of Comparative Pathology*, **100**, 419–26.

WOLF, B. C. and NEIMAN, R. S. (1989) Embryology and anatomy, in BENNINGTON, J. L. (Ed.) *Disorders of the Spleen*, pp. 3–19, Philadelphia: Saunders.

9

The Female Genital System

9.1 Ovaries

9.1.1 *Non-neoplastic Changes*

The weight and irregular shape of the rat ovary change to some extent during the oestrous cycle, so that histological sampling can be variable and single sections may not provide a good representation of the various components. If it is necessary to study ovarian morphology in detail, step serial sections are necessary.

Ovarian weights

The weight of the ovary changes slightly during the stages of the oestrous cycle, but the weights can be a useful indication of any treatment which has interfered with the complex control systems of ovarian function. Table 9.1 shows the weights of the ovaries at different time points. They increase in absolute weight from 12 to 34 weeks, but at 58 weeks of age have decreased again as function declines. Ovarian weight as a percentage of body weight declines from 12 weeks as non-lean body mass increases. These ovarian weights are similar to those reported for the SD rat and other Wistar strains (Peluso and Gordon, 1994).

Functional changes

The control of ovarian function in the rat is dependant on a complex inter-relationship between the ovaries, endocrine and nervous system. Spontaneous changes in the ovaries are likely, therefore, to be due to alterations in one or

Table 9.1 Ovarian weights in the AP rat

Age (weeks)	Number	Ovarian weights	
		Absolute weight (g)	Relative weight (% of body weight)
12	10	0.103	0.045
34	20	0.109	0.033
58	25	0.104	0.027

more of the various organs involved in controlling ovarian function. The hypothalamus is of pivotal importance in the regulation of ovarian function, as it secretes the gonadotrophin releasing hormones (GnRH) which control release of follicular stimulating hormone (FSH) and luteinising hormone (LH) from the pituitary gland. It is usual to ascribe the control of follicular development to FSH and ovulation to LH; these hormones are also involved in the negative feedback to the pituitary and hypothalamus. The secretion of GnRH can be affected by a range of factors which send stimuli to the hypothalamus from other areas of the brain. Among these factors are stress and olfactory stimuli such as pheromones. Light/dark periods can also affect secretion. Beys *et al.* (1995) showed that ovarian atrophy developed in SD rats exposed to low intensity light during the nocturnal 12 hours. Continuous exposure to light, of daylight intensity, disrupts the oestrous cycle (Campbell and Schwartz, 1980) by lowering the levels of FSH, LH and progesterone (Takeo, 1984). In addition to GnRH, FSH and LH, the ovaries produce oestradiol from the follicular granulosa cells, progesterone from the corpus luteum, and androgens and progesterone from the interstitial cells. These hormones also have regulatory controls and are involved in feedback mechanisms to the pituitary and hypothalamus, as well as other activities within the reproductive system. Apart from their steroid hormone functions, interstitial cells have been shown recently to produce transforming growth factor β, growth factor a and epidermal growth factor (Kudlow *et al.*, 1987; Skinner *et al.*, 1987; Bendell and Dorrington, 1988). This indicates that the control of ovarian function has yet to be fully elucidated, and growth factors may have an important role.

Oestrus cycle

During the normal oestrous cycle the egg-containing (primordial) follicles begin to grow, and some secrete oestradiol; at oestrus, between five and six follicles have matured and ovulate, each releasing an oocyte into the Fallopian tubes. The follicle which has shed its ovum then differentiates into a corpus luteum which secretes progesterone. Most of the follicles which did not ovulate undergo a degenerative process known as atresia, and finally form interstitial cells. A few do not ovulate or undergo atresia, but form cysts. Thus, in AP rats up to 6 months of age, the ovary may show the following changes: follicles in all stages of development, ranging from a primordial follicle with a single layer of granulosa cells around it, to a mature Graafian follicle with a fluid-filled antrum; atretic follicles in various stages of degeneration are present, some with degenerating ova, others in the final stage of interstitial cell formation. Eight to ten corpora lutea are present, including those newly formed and those regressing at the end of a cycle (Figure 74). In the AP rat at 18 months the females show a marked decline in ovarian function; the ovaries are smaller with absent or infrequent corpora lutea and few Graafian follicles. There are few follicular cysts and developing follicles (Figure 75). Finally the ovaries become small and atrophic with no corpora lutea or mature follicles and variable numbers of interstitial cells. Forty per cent of AP females show this stage of almost total atrophy in 2 year studies, but in the life-span study several females over 48 months still showed the presence of immature follicles. Low levels of ovarian atrophy have been reported in other strains, such as a 0.2 per cent incidence in 2 year studies in F344 rats (Montgomery and Alison, 1987). It seems unlikely that the F344 rat retains normal ovarian function to this age and that the low incidence is a failure to record atrophy, which many regard as a normal change in the old animal.

The physiology of the rat ovary has been described in detail by Peluso (1994). The duration of the oestrous cycle is 4 days in the AP rat, but varies

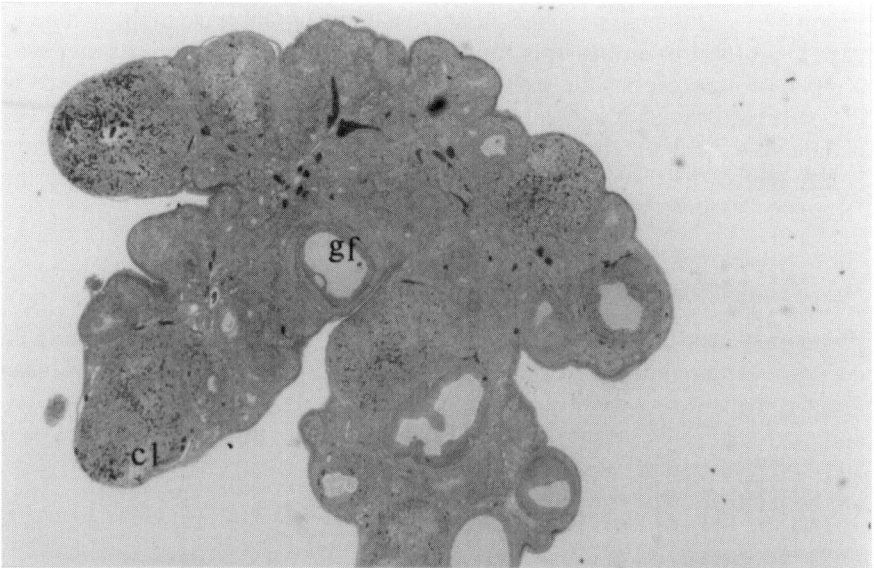

Figure 74 Normal ovary from an AP rat aged 6 months: follicles of all stages of development are present, including Graafian follicles (gf), and there are several corpora lutea (cl). ×8, H&E

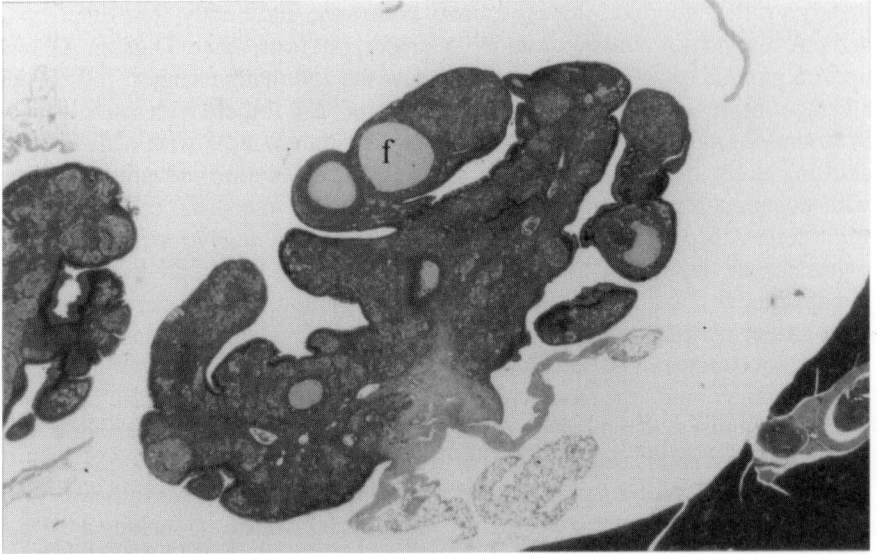

Figure 75 Ovarian atrophy in an AP rat aged 18 months: there are no corpora lutea or mature follicles. Several follicular cysts (f) are present. ×8, H&E

between 4 and 5 days in other strains (Long and Evans, 1922). The cycle becomes longer and more irregular with age, and at 12 months in the AP rat it is 5 to 6 days in duration. This has been reported for other strains (Butcher and Page, 1981). The increase in length may be due to an increase in the number of days in pro-oestrus or dioestrus. When oestrous cycles cease, the rats may enter a phase of constant oestrus where the ovarian morphology is characterised by the presence of numerous fluid-filled follicular cysts (Steger *et al.*, 1976). This is thought to be due to the extended pro-oestrus phase in older animals (Peluso and England-Charlesworth, 1981). The final anoestrous phase appears around 18 months of age in the majority of AP rats.

Pseudopregnancy

Pseudopregancy in the AP rat, characterised by large persistent corpora lutea, occurs in approximately 2 per cent of AP females over 12 months of age. Pseudopregnant rats have high progesterone levels because of the persistent corpora lutea. As in most rats over 12 months of age, prolactin levels are high, and this may be the cause of the persistence of the corpora lutea as prolactin has both luteotrophic and luteolytic activity.

Food restriction

Restricting food intake prolongs reproductive function in the rat (Merry and Holehan, 1979). Ovulation and litter size are reduced, and thus the reduction in the loss of oocytes accounts for the prolongation of fertility. Meredith *et al.* (1986) suggested that food restricted rats have a decreased sensitivity to FSH since this hormone is elevated while LH is suppressed, but it may be the change in the FSH/LH ratio which prevents normal function. A protein deficient diet causes ovarian atrophy, and low fat diets alter cycles and suppress ovulation. These effects have important implications for carcino-genicity studies where a reduction of body weight, in the group treated with the test material, is a common criterion for dose selection. As many diets have a high protein level, food restriction is most likely to affect this component.

Cystic bursa/cystic rete tubules

Cystic dilatation of the ovarian bursa is uncommon (2 per cent) in the AP rat, and it is often difficult to detect in sections since the bursa may collapse if perforated at necropsy. It should be distinguished from ovarian cysts and from the paraovarian cysts which are found infrequently in the mesovarium. These are vestigial remnants of embryonic ducts. Montgomery and Alison (1987) recorded 131 paraovarian cysts in 11 444 F344 rats. The incidence is less than 1 per cent in the AP rat. Cystic rete tubules occur in the hilus of the ovary (Figure 76) and differ from other types of ovarian cyst. Unlike follicular cysts granulosa cells are absent in cystic rete tubules, and there is no smooth muscle in the wall as occurs in paraovarian cysts of the mesovarium. The derivation of these structures is uncertain.

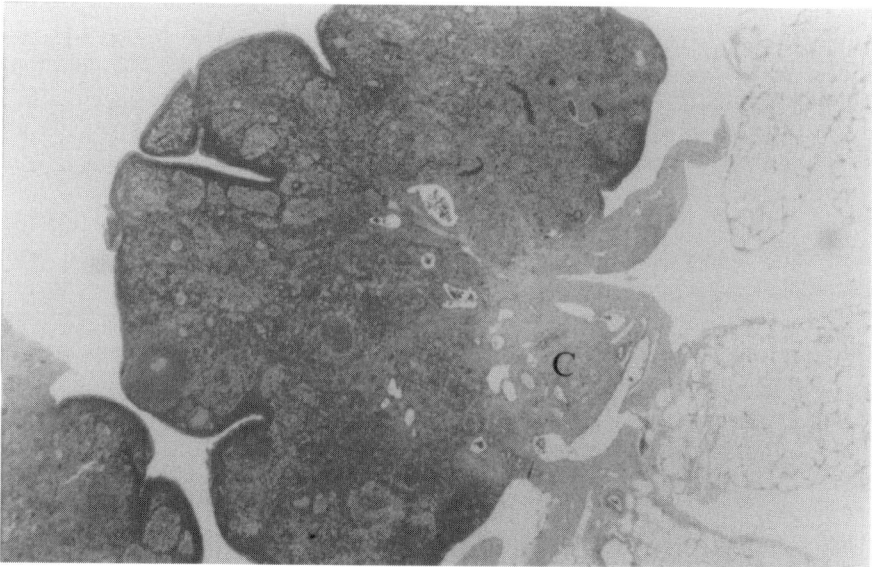

Figure 76   Cystic rete tubules in the ovarian hilus (C). ×8, H&E

Inflammation

Inflammation of the ovary is a very rare condition in virgin animals (Montgomery and Alison, 1987); in the AP rat database it has only been observed in three animals as an ascending infection from the uterus. In breeding rats, where uterine infections are more common, the incidence remains less than 1 per cent.

Pigmentation

Ceroid (lipofuscin) pigment is common in the ovaries of old rats, including the AP rat, and has been reported in the F344 rat by Montgomery and Alison (1987).

Arteritis

Arteritis is uncommon in the ovary (<1 per cent) even when the disease is widespread in other tissues. An incidence of 0.05 per cent in the F344 rat was recorded by Montgomery and Alison (1987).

Hyperplasia

Stromal hyperplasia occurs in atrophic ovaries in approximately 2 per cent of AP rats. The ovaries may be enlarged with masses of large, vacuolated interstitial cells, some in tubular or follicular groups. The mechanism for this stromal hyperplasia is likely to be lack of the negative feedback to the pituitary. In the atrophic ovary, oestrogen production declines and the lack of this negative feedback increases the pituitary secretion of FSH and LH. As interstitial cells have LH receptors the increased level of this hormone stimulates the interstitial cells to proliferate.

Small nests of Sertoli-like tubular hyperplasia are an infrequent occurrence in the ovaries of older animals, particularly in atrophic ovaries and those with granulosa cell tumours. These small areas show tubular structures lined by cells which resemble Sertoli cells. Similar changes were described by Burek (1978). The tubules are thought to be the result of follicular degeneration with granulosa cells converting into Sertoli cells (Crumeyrolle-Arias and Aschheim, 1981). They should be distinguished from sertoliform tubular adenomas. Similar changes can be induced in the rat by hypophysectomy and growth hormone (Gopinath *et al.*, 1988).

9.1.2 Neoplastic Changes

The incidence of ovarian tumours in 2 year studies with the AP rat varies between 2 and 6 per cent. They are mostly small tumours which occur as incidental findings at necropsy. Tumours of granulosa cell origin account for the majority of ovarian tumours in AP rats, unlike humans where most ovarian tumours are of epithelial origin, and sex cord tumours are rare. This low incidence of ovarian tumours has been reported for all strains of rat (MacKenzie and Garner, 1973; Carter and Ird, 1976; Lang, 1986; Lewis, 1987; Maekawa and Hayashi, 1987; Alison and Morgan, 1987).

The histological types of ovarian tumour which have been seen in the AP rat are shown in Table 9.2. The incidence of granulosa cell tumours varies between 0 and 4 per cent in 2 year studies, but the incidence of all other types is less than 1 per cent.

Sex cord tumours

The majority of sex cord tumours in the AP rat have been a mixture of granulosa, thecal and luteinised cells, and the classification of the tumour is based on the dominant cell type, usually a granulosa cell. Approximately one third of the tumours were bilateral. The architecture of the tumours may be in sheets, follicles or cords; the cells are small and basophilic with few mitotic

Table 9.2 Histological types of ovarian tumour in the AP rat

Histological type of tumour	Number observed[a]
Epithelial tumours	
Cystadenoma	3
Sertoliform tubular adenoma	3
Adenocarcinoma	1
Sex cord tumours	
Granulosa cell	48
Thecal cell	2
Luteal cell	2
Sertoli cell	3
Germ cell tumours	
Choriocarcinoma	1
Dysgerminoma	1
Teratoma	1
Others	
Fibroma	1
Angioma	1
Mesothelioma	2

[a] The number observed in a database of 4338 female AP rats (including 2500 females in 2 year studies) used in toxicology studies between 1960 and 1994.

figures. Thecal cells are spindle shaped and usually in whorls or small bundles. Only a few tumours have been composed predominantly of thecal cells, and fewer have been diagnosed as malignant granulosa cell tumours on the basis of local invasion; one had lymph node metastases. A single luteoma was distinguished from a persistent corpus luteum on its large size, pleomorphic, eosinophilic cells and mitotic figures.

Sertoli cell tumours are rare, and the three unilateral tumours seen in the AP rat were located at the hilus and were clearly demarcated from surrounding ovarian tissue (Figure 77), but were not encapsulated. They were characterised histologically by tubules lined by cells with a basal nucleus and abundant cytoplasm arranged in a perpendicular position to the basement membrane of the tubule (Figure 78). A minimal stroma was present between the tubules. These sertoli cell tumours were not locally invasive and did not metastasise. The derivation of these tumours is not certain. It has been suggested that they may be derived from the rete ovarii which has the same embryonic origin as the testis (Stoica *et al.*, 1987).

Epithelial tumours include cystadenomas where the tumour has a large cyst and a small area of tubular differentiation. Sertoliform tubular adenomas were composed of well-defined tubules lined by sertoli-like cells without the clear basal nuclei and vertical arrangement of the sertoli cell tumour. A variable stroma of interstitial cells was also present (Figure 79). The adenocarcinoma was a well-differentiated tumour of tubular structures, which was locally invasive.

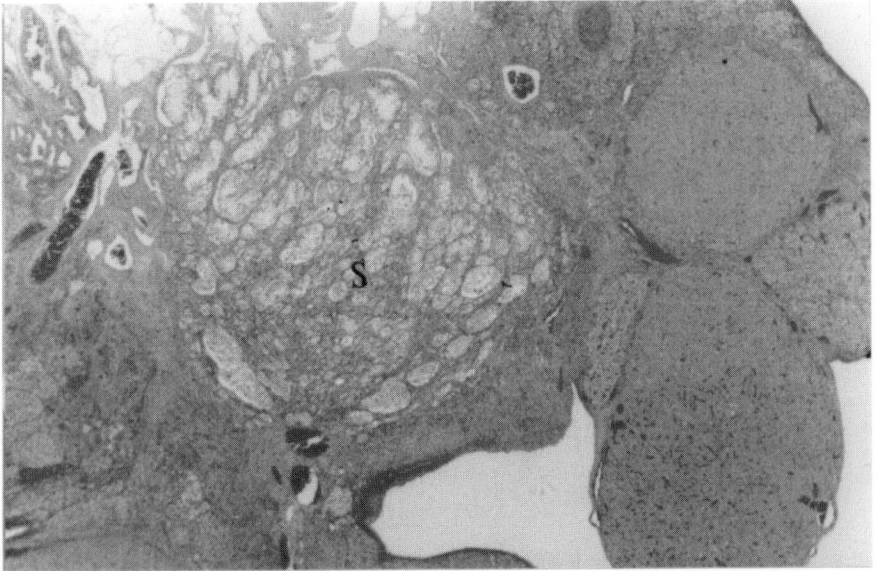

Figure 77 Sertoli cell tumour (S) in the ovary of an AP rat; the tumour is clearly demarcated from surrounding tissue. ×32, H&E

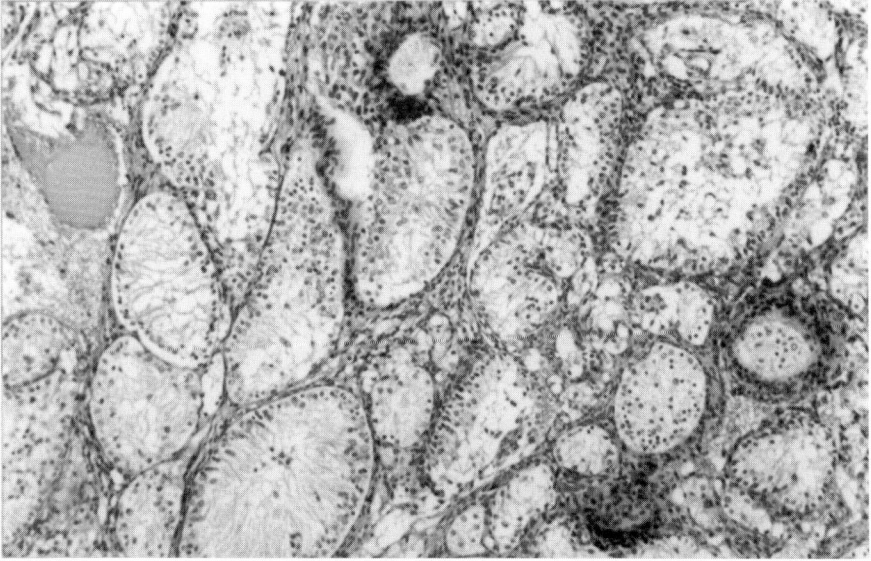

Figure 78 Sertoli cell tumour showing the characteristic tubular structures lined by Sertoli-like cells arranged in a perpendicular position to the basement membrane. ×80, H&E

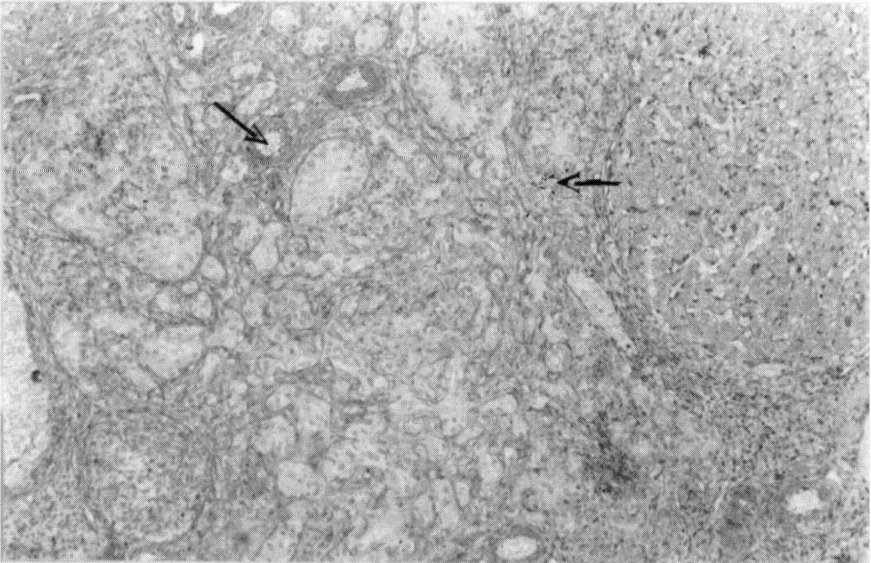

Figure 79 Sertoliform tubular adenoma (←) in an AP rat: the tumour is not clearly differentiated from the surrounding tissue, and the tubular structures are lined by cells without the perpendicular arrangement of the Sertoli cell tumour. ×80, H&E

153

The single example of choriocarcinoma was a haemorrhagic tumour of cytotrophoblastic cells with very large nuclei. Metastases were found in the lungs. The unilateral dysgerminoma was a very large tumour, the size of a walnut and showed cords of large epithelial cells, separated by a scant fibrous stroma, with a few areas of lymphocyte infiltration. The single unilateral teratoma contained neurological, pulmonary and connective tissue.

The miscellaneous tumours, fibroma and angioma, had the histological appearance of the tumours in the skin while the mesotheliomas had a papillary arrangement and had spread on the peritoneal surfaces.

9.2 Uterus

The uterus of the rat has two cornua (horns) which are widely separated cranially, but caudally are close together. Where it opens into the vagina there are several folds, the portio vaginalis. In the AP rat for many years only the left horn of the uterus was taken for histological examination; the remainder of the tract was opened and inspected. Latterly both horns have been taken and a sample through the caudal area of both horns taken for histological examination, and an additional sample through the cervix. The remainder of the uterus is opened and inspected. Absence of one or both uterine horns has been seen rarely.

9.2.1 *Non-neoplastic Changes*

Uterine weights

The weights of the uterus at various ages are shown in Table 9.3 and follow the pattern for the weights of other organs. There is an increase in both absolute and relative uterine weights between 32 and 56 weeks as the animals attain maximum lean body weight. Thereafter there is little change in the absolute weight and the relative weight declines.

Table 9.3 Uterine weights in the AP rat

Time (weeks)	Number	Mean uterine weight	
		Absolute weight (g)	Relative weight (% body weight)
12	10	0.29	0.12
32	20	0.58	0.17
56	25	0.59	0.15

Functional changes

The histological appearance of the uterus in the mature animal varies according to the stage of the oestrous cycle but the stages of the cycle are identified more readily in the vagina. In the proestrous phase which lasts 12 hours the uterine horns are dilated with fluid, and blood vessels are congested. Oestrus, which also lasts 12 hours, is characterised by the continuing presence of luminal dilatation, although fluid is lost in the first few hours as the cervix relaxes; there is a little vacuolation of the lining epithelium, and neutrophils collect around glands, and below and within the lining epithelium. In metoestrous, which lasts 20 to 21 hours, there is a more marked vacuolation and degeneration of the epithelium. In dioestrous, the longest phase at 48–57 hours, the lining epithelium of the endometrium returns to the resting, simple columnar epithelium. In the endometrium, scattered glands are present, and eosinophils are often quite numerous in the loose connective tissue of the lamina propria, at all stages of the cycle. In the state of constant oestrus, which marks the end of reproductive life, there is an increase in collagen in the endometrium; finally in the stage of persistent dioestrous the endometrium has a dense collagen stroma and few inactive glands.

Inflammation

Endometritis and pyometra are rare conditions in the virgin AP rat and chiefly occur when there are tumours in the uterus. The cyclical increase in neutrophils in the endometrium can be mistakenly diagnosed as mild acute endometritis, and this diagnosis should not be used in the absence of any other change. Inflammatory conditions of the uterus are more frequent in breeding females (>5 per cent).

Pigmentation/vascular disease

Pigment-laden macrophages accumulate with age in the endometrium, and in very old rats significant quantities of lipofuscin may be present. Polyarteritis is rare in the uterus of the AP rat, and in animals with widespread arterial disease only 5 per cent show uterine involvement.

Dilatation/hyperplasia

The term luminal dilatation or hydrometra is often recorded in rats, and is most likely to be due to the normal cyclical changes in the uterus (Leininger and Jokinen, 1990). Cystic endometrial hyperplasia (Figure 80) occurs in around 2 per cent of AP rats over 18 months of age. This is a proliferative lesion with increased mitotic activity in the glands, which may be multilayered. By contrast, cystic endometrial glands is the term used to describe one or two cystically dilated glands in the endometrium; it is found in 6 per cent of animals in 2 year studies.

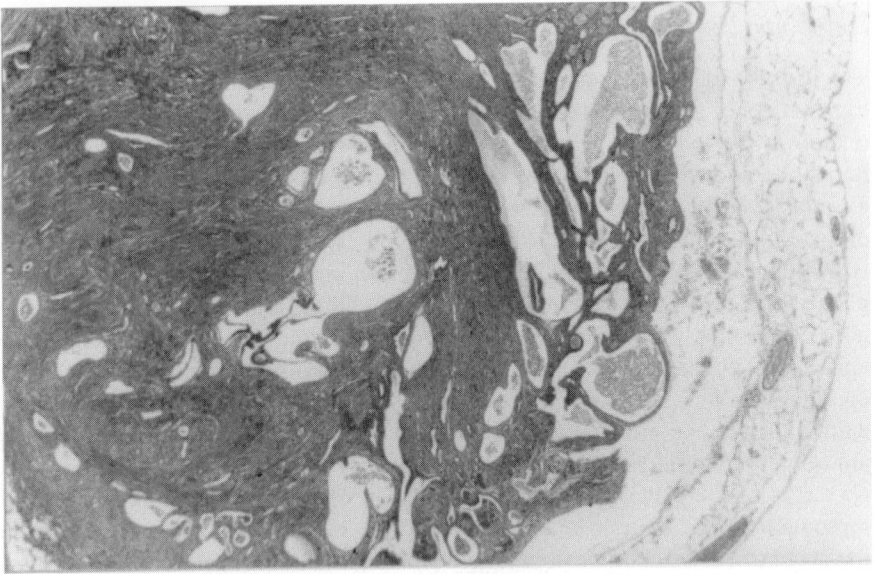

Figure 80 Cystic endometrial hyperplasia in a 22 month old AP rat. ×32, H&E

Squamous metaplasia of glands usually occurs in single glands when there is cystic hyperplasia or when there is an inflammatory condition in the uterus.

Decidual reaction

Decidual reaction (deciduoma) is a rare condition in most strains (Elcock *et al.*, 1987). It arises in the uterus of the non-pregnant rat, and shows large round stromal cells with an extensive eosinophilic cytoplasm and large nuclei containing prominent nucleoli. Three AP rats less than 6 months of age have shown decidual reactions in the uterus. Ohta (1987) showed that there was a decline in the induction of decidual reactions with time. The hormonal requirements for deciduoma include elevation of progesterone level for at least 2 days, followed by an increase in oestrogen. Mechanical irritation is also a possible factor.

Fallopian tube/cervix

Few changes have been observed in the fallopian tubes as they are not examined routinely, only when macroscopically abnormal or when present in sections of the ovaries. Salpingitis has been observed as an ascending infection and cystic dilatation is an uncommon finding.

The cervix has shown a congenital abnormality of structures resembling primordial ovarian follicles in the myometrium in three animals. Cervicitis is very rare and secondary to vaginal or uterine infections.

9.1.2 Neoplastic Changes

Cervix/Fallopian tubes

A single papilloma has been observed in the cervix, and a leiomyoma in the oviduct.

Uterus

Tumours of the uterus are among the most common in the AP rat. The incidence of various types is shown in Table 9.4. Stromal endometrial polyps are the most common type and may be sessile or, more frequently, pedunculated. They show a range of histological forms. In the AP rat the most common appearance is of a dense stroma containing numerous capillaries with a lining epithelium of flat or cuboidal cells around the polyp (Figure 81). This type of polyp is sometimes called a vascular polyp. Glands are absent from this type of polyp, which distinguishes them from the small number of polyps which do show large numbers of glands and are classified as adenomatous polyps/adenomas. The adenocarcinomas have all been poorly differentiated tumours of columnar or cuboidal epithelium in mixed glandular and papillary patterns. A scanty stroma was present and there was considerable mitotic activity. All adenocarcinomas were associated with extensive inflammatory changes and showed invasion of the myometrium and tissues adjacent to the uterus. Lung metastases have been observed with two tumours. The histological appearances of all other tumours are similar to these tumours at other sites. In the AP rat uterine tumours are more common in older animals. The majority of polyps were incidental findings and all other tumours were found in animals over 20 months of age. The incidence of endometrial polyps in AP rats is higher in later studies than in those completed between 1960 and 1980. This can be attributed to the increased histological sampling of the uterus in later years. Uterine tumours in other strains show

Table 9.4 Incidence of uterine tumours in the AP rat

Type of tumour	Highest incidence observed[a] (%)
Endometrial stromal polyp	14.0
Adenomatous polyp (adenoma)	1.0
Adenocarcinoma	1.0
Leiomyoma	1.0
Leiomyosarcoma	2.0
Fibrosarcoma	1.5
Angioma	1.5
Fibrohistiocytic sarcoma	1.3

[a] Incidence in 4338 female AP rats (including 2500 used in 2 year studies) used in toxicology studies between 1960 and 1992.

157

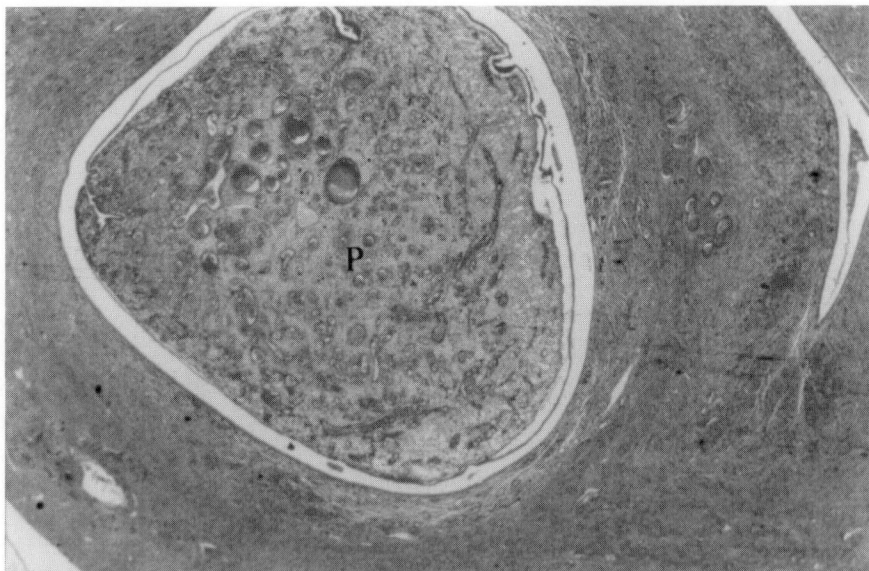

Figure 81 Stromal polyp (P) in the uterus of a 21 month old rat. ×8, H&E

considerable variation. The Han:Wistar is reported to have a 5 per cent incidence of uterine adenocarcinoma at two years but this rises to 39 per cent in a life-span study, with 40 per cent lung metastases (Deerberg *et al.*, 1981). Kroes *et al.* (1981) reported a total uterine tumour incidence of less than 1 per cent in another Wistar strain. In SD rats low incidences of less than 2 per cent have been reported by Muraoka *et al.* (1977) with 60 per cent occurring after 2 years. The marked increase in tumours after 2 years was also seen in BN/Bi rats (Burek, 1978). Endometrial stromal polyps are the most common tumour of the F344 rats, with low incidences of adenocarcinomas in 2 year studies (Maekawa *et al.*, 1983; Sollveld *et al.*, 1984; Goodman and Hildebrandt, 1987), but again in 31 month old F344 rats a 55 per cent incidence of adenocarcinoma has been reported by Tang and Tang (1981). In the life-span study in the AP rat the incidence of adenocarcinomas of the uterus was 4/137 (2.9 per cent) in virgin rats and 4/151 (2.6 per cent) in breeding females, and all of the tumours were seen in animals over 28 months of age.

9.3 Vagina

9.3.1 *Non-neoplastic Changes*

The vagina has only been examined routinely in the last 15 years in AP rats and, other than functional changes, few non-neoplastic conditions have been observed. The vagina provides the best means of assessing the stage of the oestrous cycle. In histological sections the vaginal epithelium in the dioestrous phase has between eight and ten cell layers infiltrated by leukocytes; in proestrous the superficial layers show mucous vacuolation and deeper layers become keratinised, and at oestrus the epithelium has six to ten cell layers with cornified superficial layers. In metoestrous the superficial layers slough off into the vaginal lumen until dioestrous begins and the cell layers increase again. In persistent oestrous at the end of reproductive life the vaginal epithelium remains keratinised, and, in the persistent dioestrous which follows, some or possibly all layers of the epithelium show mucous vacuolation.

Imperforate vagina is an uncommon condition affecting less than 1 per cent of AP rats. Affected animals usually develop a swollen lower abdomen within the first few months and are removed from the study. The vagina is dilated and there is usually some inflammation of the genital tract cranial to the vagina. Vaginal septa have been observed in other Wistar rats (de Schaepdrijver *et al.*, 1995).

Vaginal inflammation is a rare condition and only mild chronic inflammation has been observed.

9.3.2 Neoplastic Conditions

The incidence of vaginal tumours is low in the AP rat in common with most strains of rat, except the Brown Norway (Burek *et al.*, 1978), with an overall incidence in carcinogenicity studies of less than 1 per cent. Only single tumours have been found in any of the 24 carcinogenicity studies. The tumours include a squamous carcinoma, two fibrosarcomas and four fibrohistiocytic sarcomas.

9.4 References

ALISON, R. H. and MORGAN, K. T. (1987) Ovarian neoplasms in F344 rats and B6C3F1 mice, *Environmental Health Perspectives*, **73**, 91–106.

BENDELL, J. J. and DORRINGTON, J. (1988) Rat thecal/interstitial cells secrete a transforming growth factor-β-like factor that promotes growth and differentiation in rat granulosa cells, *Endocrinology*, **123**, 941–8.

BEYS, E., HODGE, T. and NOHYNEK, G. J. (1995) Ovarian changes in Sprague-Dawley rats produced by nocturnal exposure to low intensity light, *Laboratory Animal Science*, **29**, 335–8.

BUREK, J. D. (1978) *Pathology of the Aging Rat*, pp. 117–23, West Palm Beach, Florida: CRC Press.

BUREK, J. D., ZURCHER, C. and HOLLANDER, C. F. (1978) High incidence of spontaneous cervical and vaginal tumors in an inbred strain of Brown Norway rats, *Journal of the National Cancer Institute*, **57**, 549–54.

BUTCHER, R. L. and PAGE, R. D. (1981) Role of the aging ovary in cessation of reproduction, in SCHEARTZ, N. B. and HUNZICKER-DUNN, M. (Eds), *Dynamics of Ovarian Function*, pp. 611–15, New York: Raven Press.

CAMPBELL, C. S. and SCHWARTZ, N. B. (1980) The impact of constant light on the estrus cycle of the rat, *Endocrinology*, **106**, 1230–8.

CARTER, R. L. and IRD, E. A. (1976) Tumours of the ovary, in TURUSOV, V. S. (Ed.), *Pathology of Tumours in Laboratory Animals*, Vol. 1, *Tumours of the Rat*, pp. 189–200, Lyon: IARC.

CRUMEYROLLE-ARIAS, M. and ASCHHEIM, P. (1981) Post-hypophysectomy ovarian senescence and its relation to the spontaneous structural changes in the ovary of intact aged rats, *Gerontology*, **27**, 58–71.

DEERBERG, F., REHM, S. and PITTERMAN, W. (1981) Uncommon frequency of adenocarcinomas of the uterus in virgin Han:Wistar rats, *Veterinary Pathology*, **18**, 707–13.

ELCOCK, L. H., STUART, B. P., MUELLER, R. E. and MOSS, H. E. (1987) Deciduoma, uterus, rat, in JONES, T. C., MOHR, U. and HUNT, R. D. (Eds), *Genital System*, pp. 140–6, Berlin and Heidelberg: Springer-Verlag.

GOODMAN, D. G. and HILDEBRANDT, P. K. (1987) Stromal polyp, endometrium, rat, in JONES, T. C., MOHR, U. and HUNT, R. D. (Eds), *Genital System*, pp. 146–8, Berlin and Heidelberg: Springer-Verlag.

GOPINATH, C., PRENTICE, D. E. and LEWIS, D. J. (1988) *Atlas of Toxicological Pathology*, pp. 91–103, Norwall, Massachusetts: MTP Press.

KROES, R., GARBIS-BERKVENS, J. M., DE VRIES, T. and VAN NESSELROOY, J. H. J. (1981) Histopathological profile of a Wistar rat stock including a survey of the literature, *Journal of Gerontology*, **36**, 259–79.

KUDLOW, J. E., KOBRIN, M. S., PURCHIO, A. F., TWARDZIK, D. R., HERNANDEZ, E. R., ASA, S. L., SKINNER, M. K., LOBB, D. and DORRINGTON, J. H. (1987) Ovarian thecal/interstitial cells produce an epidermal growth factor like substance, *Endocrinology*, **121**, 1892–9.

LANG, P. L. (1986) *Spontaneous Neoplastic Lesions in the Crl:CD BR Rat*, pp. 1–13, Kingston, New York: Charles River Laboratories.

LEININGER, J. R. and JOKINEN, M. P. (1990) Oviduct, uterus and vagina, in BOORMAN, G. A., EUSTIS, S. L., ELWELL, M. R., MONTGOMERY, C. A. and MACKENZIE, W. F. (Eds), *Pathology of the Fischer Rat*, pp. 443–59, San Diego: Academic Press.

LEWIS, D. J. (1987) Ovarian neoplasia in the Sprague-Dawley rat, *Environmental Health Perspectives*, **73**, 77–90.

LONG, J. A. and EVANS, H. M. (1922) The estrus cycle in the rat and its associated phenomena, *Memoirs of the University of California*, Vol. 6, Berkley: University of California Press.

MACKENZIE, W. F. and GARNER, F. M. (1973) Comparison of neoplasms in six sources of rats, *Journal of the National Cancer Institute*, **50**, 1243–57.

MAEKAWA, A. and HAYASHI, Y. (1987) Granulosa/thexa cell tumor, in JONES, T. C., MOHR, U. and HUNT, R. D. (Eds), *Genital System*, pp. 15–22, Berlin and Heidelberg: Springer-Verlag.

MAEKAWA, A., KUROKAWA, Y., TAKAHASHI, M., KOKUBO, T., OGIU, T., ONODERA, H., TANIGAWA, H., OHNO, Y., FURUKAWA, F. and HAYASSHI, Y. (1983) Spontaneous tumours in F344/DuCrj rats, *Gann*, **74**, 365–72.

MEREDITH, S., KIRKPATRICK, K. and BUTCHER, R. L. (1986) The effects of food restriction and hypophysectomy on numbers of primordial follicles and concentrations of hormones in rats, *Biology of Reproduction*, **35**, 68–73.

MERRY, B. J. and HOLEHAN, A. M. (1979) Onset of puberty and duration of fertility in rats fed a restricted diet, *Journal of Reproduction and Fertility*, **57**, 253–9.

MONTGOMERY, C. A. and ALISON, R. H. (1987) Non-neoplastic lesions of the ovary in Fischer 344 rats and B6C3F1 mice, *Environmental Health Perspectives*, **73**, 53–75.

MURAOKA, Y., ITOH, M., MAEDA, Y. and HAYASHI, Y. (1977) Histological changes of various organs in aged SD-JCL rats, *Experimental Animals (Japan)*, **26**, 13–22.

OHTA, Y. (1987) Age-related decline in deciduogenic ability of the rat uterus, *Biology of Reproduction*, **37**, 779–85.

PELUSO, J. J. (1994) Morphologic and physiologic features of the ovary, in MOHR, U., DUNGWORTH, D. L. and CAPEN, C. C. (Eds), *Pathobiology of the Aging Rat*, Vol. 1., pp. 338–49, Washington: ILSI Press.

PELUSO, J. J. and ENGLAND-CHARLESWORTH, C. (1981) Formation of ovarian cysts in aged, irregularly cycling rats, *Biology and Reproduction*, **24**, 1183–90.

PELUSO, J. J. and GORDON, L. R. (1994) Non-neoplastic and neoplastic changes in the ovary, in MOHR, U., DUNGWORTH, D. L. and CAPEN, C. C. (Eds), *Pathobiology of the Aging Rat*, Vol. 1, pp. 353–64, Washington: ILSI Press.

DE SCHAEPDRIJVER, L. M., FRANSEN, J. L., VAN DER EYCKEN, E. S. and COUSSEMENT, W. C. (1995) Transverse vaginal septum in specific pathogen free Wistar rat, *Laboratory Animal Science*, **45**(2), 181–3.

SKINNER, M. K., LOBB, D. and DORRINGTON, J. H. (1987) Ovarian thecal/interstitial cells produce an epidermal growth factor-like substance, *Endocrinology*, **121**, 1892–9.

SOLLVELD, H. A., HASEMAN, J. K. and MCCONNELL, E. E. (1984) Natural history of body weight gain, survival and neoplasia in the F344 rat, *Journal of the National Cancer Institute*, **72**, 929–40.

STEGER, R. W., PELUSO, J. J., HUANG, J., MEITES, J. and HAFEZ, E. (1976) Gonadotropin binding sites in the ovary of aged rats, *Journal of Reproduction and Fertility*, **48**, 205–7.

STOICA, G., CAPEN, C. C. and KOESTNER, A. (1987), in JONES, T. C., MOHR, U. and HUNT, R. D. (Eds), *Genital System*, pp. 30–6, Berlin and Heidelberg: Springer-Verlag.

TAKEO, Y. (1984) Influences of continuous illumination on the estrus cycle of rats: time course of changes in levels of gonadotrophins and ovarian steroids until occurrence of persistent estrus, *Neuroendocrinology*, **39**, 97–104.

TANG, F. Y. and TANG, L. K. (1981) Association of endometrial tumors with reproductive tract abnormalities in the aged rat, *Gynecologic Oncology*, **12**, 51–63.

10

The Male Genital System

Both testes and epididymides, ventral prostate and the left seminal vesicle are examined routinely in all toxicology studies with the AP rat. In the majority of studies the testes have been immersion fixed in buffered formalin, after cutting the capsule, but latterly Bouin's fixative has been used to allow tubular staging when this was considered necessary. The problems of fixation of the testes have been detailed by Lamb and Chapin (1985).

10.1 Testes

10.1.1 Non-neoplastic Changes

Testicular weights

The weight of the testis in AP rats, at different time points, is shown in Table 10.1. The absolute weight of the testes increases up to 1 year and then

Table 10.1 Testicular weights in the AP rat

Age in weeks	Number	Mean testicular weight[a]	
		Absolute weight (g)	Relative weight (% of body weight)
12	10	4.04	1.16
32	20	4.84	0.79
58	25	5.25	0.72
108	50	4.26	0.71

[a] Mean weight of testes and epididymides.

decreases as circulating hormone levels decline and tubular atrophy develops (Chan *et al.*, 1977; Bethea and Walker, 1979). The percentage weight declines throughout the life of the rat as the body weight increases, since the testes of the adult rat, unlike most other organs, do not grow in relation to body growth. The weights are similar to those described for SD rats, other Wistar strains and the Holtzman rat (Mills *et al.*, 1977; Robb *et al.*, 1978; Russell, 1994). The exception is the relative weight of the testes in F344 rats which increases with age due to the development of interstitial cell tumours.

Functional changes

The testes of the rat descend from the abdomen into the scrotum around 15 days after birth, but the first wave of spermatogenesis begins about day 7. Sperm are produced by day 45, but optimal production occurs around postnatal day 75. Although sperm is present in the epididymis by day 50, it takes about 8 days for sperm to pass through the epididymis (Amann, 1986). Fertility in albino rats is said to be attained around 62 days (Clegg, 1960). The convoluted loops of the twenty or so seminiferous tubules are connected at both ends to the rete testis. As the duration of each step in the process of spermatogenesis is constant, it is possible to identify the stage of spermatogenesis in each tubule by the population of cells present: e.g. Figure 82 shows a tubule in stage VIII (Leblond and Clermont, 1952).

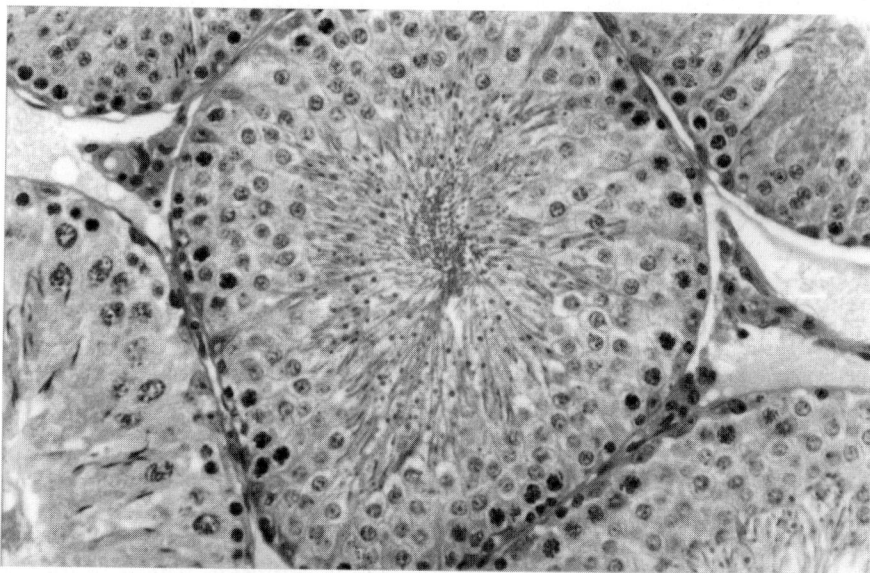

Figure 82 Seminiferous tubule in stage VIII showing elongated spermatid heads maximally curved. Basophilic granules are evident close to the spermatids. ×128, H&E

During puberty levels of luteinising hormone (LH) rise, and follicular stimulating hormone (FSH) shows a marked increase between days 12 and 21; at the end of puberty, FSH levels have declined and testosterone is the predominant androgen produced by the Leydig cells. Hormonal control of the first wave of spermatogenesis is dependant on FSH and LH; FSH binds to Sertoli cells and is critical in initiating spermatogenesis, and also in developing LH receptors on Leydig cells. The Leydig cells produce androgens under the stimulus of LH and these act on androgen receptors in the Sertoli cell. Foetal Leydig cells are replaced by adult cells during puberty; the major difference of the mature cells is that they have little lipid and rarely divide. The Leydig cell population in the adult is thought to be static as it is rare to find cells in mitosis, but they have a considerable power of regeneration (Kerr *et al.*, 1985). Their steroidogenic capacity does not decline with age (Kaler and Neaves, 1981) so that the decline in androgen levels with age is probably due to a decline in gonadotrophin levels (Gray *et al.*, 1980).

In the adult male, spermatogenesis is controlled by testosterone produced by Leydig cells under stimulation from pituitary LH. The Sertoli cells have testosterone receptors and are thought to mediate the action of testosterone on the germ cells. The role of FSH is not clear. Sharpe *et al.* (1988) consider that testosterone alone would maintain spermatogenesis, while Bartlett *et al.* (1989) consider that FSH is involved in spermatogenesis. If the levels of the essential hormones decline there is degeneration of the germ cells at specific stages of spermatogenesis (Dym *et al.*, 1977; Russell *et al.*, 1981; Sharpe, 1989). Sperm production in the rat is far in excess of that needed for fertility, and it is reported that spermatogenesis needs to be reduced between 40 and 90 per cent before fertility is impaired (Amann, 1982; Zenick and Clegg, 1989).

Atrophy

Unilateral total atrophy occurs in older animals with a maximum incidence of 10 per cent in any study. It is slightly more common in the right testis (60 per cent). In a few animals the atrophy was found in a unilateral undescended testis. In the rat the descent of the testes is thought to be a passive process due to growth of the trunk, rather than changes in the genitoinguinal ligament which connects the testes to the scrotum. The majority of males with unilateral atrophy did not have undescended testes and the cause of their condition is not known. It did not appear to be clearly correlated with any other pathological condition although in some there was polyarteritis involving both testes, and in others spermatic granulomata were present in the rete testis or epididymis.

The most common form of atrophy is an age-associated tubular atrophy. In the most severe cases the seminiferous tubule is shrunken, with total loss of germinal cells, leaving only the Sertoli cells, and slight hyperplasia of the interstitial cells of Leydig around small blood vessels (Figure 83). In the AP rat tubular atrophy can be focal or diffuse and is accompanied by a range of other

165

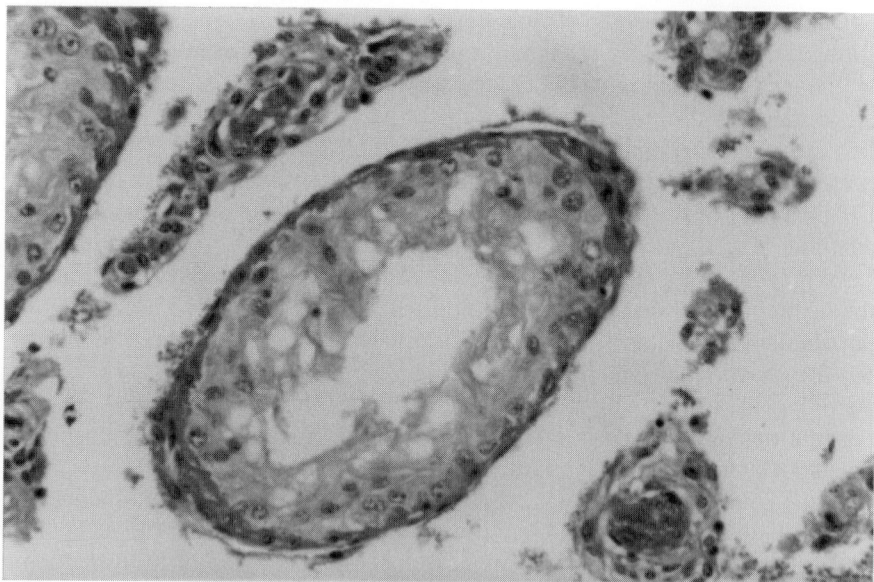

Figure 83 Atrophic seminiferous tubule in an AP rat aged 24 months: there is also hyperplasia of the interstitial cells of Leydig around small blood vessels. ×128, H&E

histological changes including vacuolation of the germ cell layer and, infrequently, multinucleate giant cells; the tubules shrink and develop thickened basement membranes and at this stage the weight of the testes is markedly reduced. Small foci of atrophic tubules may be seen in young animals of 8 weeks of age, but progression is slow and significant atrophy does not occur in the AP rat until 18 months. In 2 year studies the overall incidence of atrophy may reach 80 per cent; the majority of rats show only a focal change, but up to 10 per cent have total atrophy.

There is considerable difference in the incidence of atrophy in different strains. Burek (1978) recorded the presence of some degree of atrophy in all of his Wistar rats, and in a BN/Wistar cross, at 18 months. Wright *et al.* (1982) did not observe atrophy in their Wistar strain until day 361 and severe atrophy only at day 600. In SD rats a 20 per cent incidence of severe atrophy was present at 24 months (Heywood and James, 1985), and in F344 rats Coleman *et al.* (1977) reported a 100 per cent incidence at 18 months.

Oedema

Interstitial oedema may occur in association with a variety of pathological conditions, including tubular atrophy. It should be distinguished from the oedema which occurs in the testis as a fixation artefact (Greaves and Faccini, 1984).

Mineralisation

Tubular mineralisation occurs in tubules with severe atrophy, and the incidence may reach 25 per cent in animals in 2 year studies (Figure 84). Burek (1978) described the progression to mineralisation; tubules filled with degenerating and necrotic sperm develop small calcium deposits which increase with time as the degenerate sperm and germ cell epithelium disappear.

Spermatic granulomata

Spermatic granulomata are an infrequent finding in the AP rat (<2 per cent) and are found in dilated testicular tubules, particularly at the rete testis. The tubules contain sperm, epithelial cells and occasional granulomatous giant cells (James and Heywood, 1979).

Cystic dilatation rete testis

Cystic dilatation of the rete testis occurs when there is obstruction of the epididymal ducts. It is uncommon (<1 per cent) in the AP rat.

Polyarteritis

The testicular arteries are one of the more common sites for polyarteritis (Figure 85) in the rat and are associated with severe tubular atrophy (Greaves

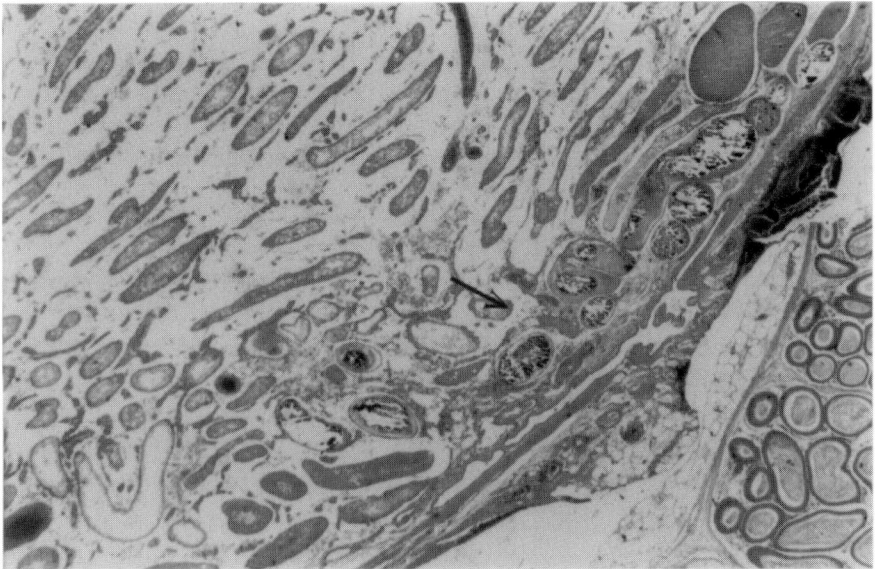

Figure 84 Subcapsular calcification of testicular tubules (→) in an atrophic testis. ×32, H&E

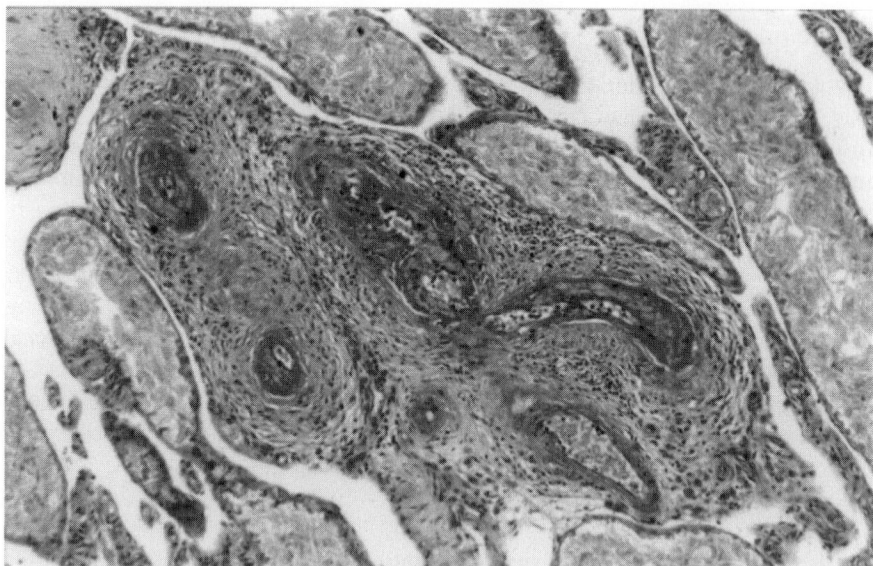

Figure 85   Polyarteritis of testicular arteries. ×80, H&E

and Faccini, 1984; James and Heywood, 1979). The incidence of polyarteritis of the testes in the AP rat reaches a maximum of 25 per cent in 2 year studies.

Hyperplasia of Leydig cells

Hyperplasia of the interstitial cells of Leydig is uncommon in the AP rat (a maximum of 3 per cent at 2 years) as in most Wistar strains. Wright *et al.* (1982) consider that there is a relationship between the hyperplasia and atrophy, rather than with the age of the animal. Thurman *et al.* (1995) showed that food restriction delays the onset of Leydig cell hyperplasia in the F344 rat. Serum testosterone is decreased in restricted animals and serum oestradiol increased. Also in the F344 rat, the incidence of hyperplasia reaches 100 per cent at 15 months although the incidence is low before 12 months (Boorman *et al.*, 1987). In the F344, hyperplasia is considered to be a pre-neoplastic change (Goodman *et al.*, 1979) and it is associated with an elevation of oestrogen and prolactin levels (Turek and Desjardins, 1979). The hyperplasia may be focal, nodular or diffuse, but the hyperplastic Leydig cells have a normal histological appearance. The distinction of hyperplasia from tumours of Leydig cells is somewhat arbitrary and the criteria are that the hyperplastic area is less than a tubule in size and shows no compression.

10.1.2   *Neoplastic Changes*

The incidence of testicular tumours varies markedly with strain but, in general,

tumours are not common in most Wistar-derived rats. The incidence in the AP
rat is shown in Table 10.2. Leydig cell tumours are the most common type;
they are usually unilateral, with only 5 per cent bilateral. In the F344 75 per
cent of tumours were reported to be bilateral (Jacobs and Huseby, 1967). They
are more common in breeding males, and the incidence increases with age, as
shown in Table 10.3 which gives the incidence in the life-span study of
breeding and non-breeding rats. There has not been any significant change in
the incidence in the AP rat in 2 year studies over the three decades the strain
has been used. Macroscopically large Leydig cell tumours have a nodular
appearance and are yellowish brown in colour. Histologically the tumours were
of variable size, often large. In the majority, the cells were uniform with round
nuclei and abundant eosinophilic cytoplasm, often vacuolated. Other cells were
intensely vacuolated, and in a few tumours the cells were small with a more
basophilic cytoplasm or occasionally spindle shaped. Mitotic figures were rare.
Cystic spaces filled with pale fluid or blood were frequent (Figure 86) but the
tumours were not encapsulated. Only one tumour has been designated
malignant; it had penetrated the capsule and invaded the adjacent epididymis,
but had not metastasised. The single example of teratoma in the right testis of a
6 month old male showed adipose tissue, muscle and pulmonary tissue, and the
fibrosarcoma which arose in the capsule showed the histological appearance of
fibrosarcomas in the skin. Mesotheliomas arise on the surface of the tunica and
have the histological appearance of numerous finger-like projections, with a
connective tissue core; these are lined by plump cells with large nuclei and
scanty cytoplasm. They may be florid growths covering the whole surface of
the testis (Figure 87). The malignant mesotheliomas were more pleomorphic
and showed widespread local invasion. Seminomas and Sertoli cell tumours
have not been found in the AP rat. Secondary tumours in the testis are also rare,
but occasional infiltration by leukaemias and lymphomas have been seen and
one metastasis of a bronchial carcinoma was found within an interstitial cell
tumour.

Table 10.2 Incidence of testicular tumours in the AP rat

Type of testicular tumour	Highest % incidence observed[a]
Benign Leydig cell tumour	15 (life-span)
	5 (2 year study)
Malignant Leydig cell tumour	1 (one tumour only)
Benign mesothelioma	2
Malignant mesothelioma	1 (two tumours only)
Teratoma	1 (one tumour only)
Fibrosarcoma capsule	1 (one tumour only)

[a] Incidence of testicular tumours in a database of 4542 male AP
rats (including 2800 males used in 2 year studies) used in
toxicology studies between 1960 and 1994.

169

Table 10.3 Incidence of Leydig cell tumours in a life-span study in the AP rat

Age of rat (months)	Number of Leydig cell tumours	
	Breeding males N = 42[a]	Non-breeding males N = 152
0–12	0	0
13–24	1	1
25–30	2	7
31–36	4	6
37–42	4	2
43–52	3	0
Total	14 (33%)	16 (10%)

[a] There were fewer breeding males as they were kept in harems (one male to four females).

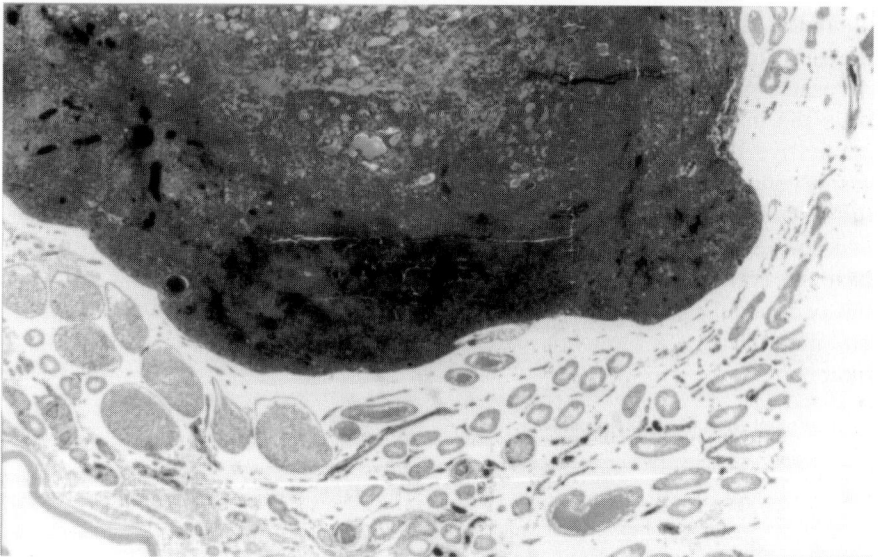

Figure 86 Leydig cell tumour in a 26 month old AP rat showing numerous blood-filled spaces. ×8, H&E

The incidence of interstitial (Leydig) cell tumours ranges from 3 per cent in some Wistar strains to 95 per cent in the F344 (Solleveld *et al.*, 1984; Bomhard *et al.*, 1986; Maita *et al.*, 1987) and they increase with age (Sass *et al.*, 1975; Coleman *et al.*, 1977; Goodman *et al.*, 1979; Maekawa *et al.*, 1983). The development of interstitial cell tumours in the F344 rat can be delayed beyond 2 years by the induction of a hyperprolactinaemic state which reduces

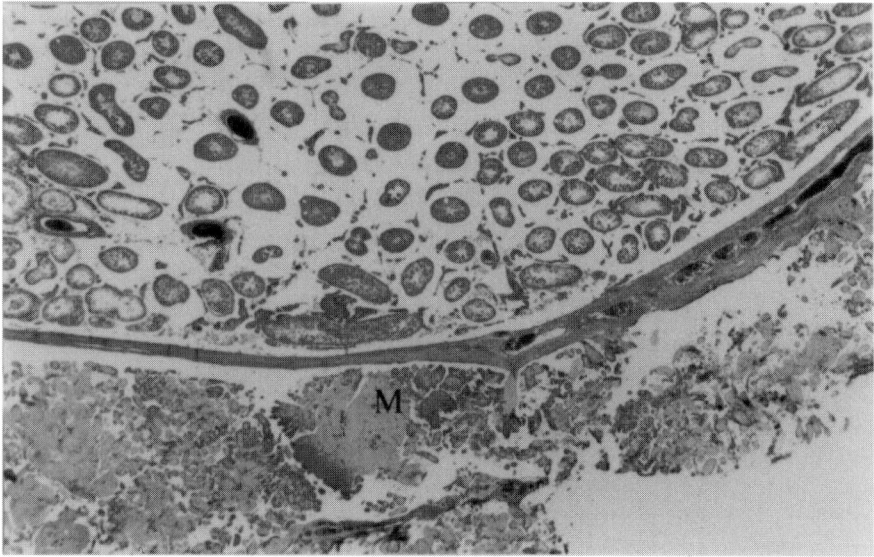

Figure 87   Mesothelioma (M) on the surface of the testis of an AP rat. ×32, H&E

peripheral luteinising hormone levels (Bartke *et al.*, 1985). This suggests that hormonal imbalance is responsible for the development of the spontaneous interstitial cell tumour. All other types of testicular tumour are rare in the rat (Sass *et al.*, 1975; Goodman *et al.*, 1980; Imai and Yoshimura, 1988). Testicular tumours have been induced in the AP rat by the hypolipidaemic drug methyl clofenapate (Tucker and Orton, 1995).

10.2   Epididymides

10.2.1   *Non-neoplastic Changes*

The epididymides are attached to the testes by a fine mesentery, and in the AP rat both epididymides are taken for histological examination; the samples include the caput, corpus and cauda epididymidis.

   In the AP rat inflammation of the epididymis is usually minimal, consisting of small mononuclear cell infiltrates which are either perivascular or interstitial. It is a common condition in the AP rat at all ages. Spermatic granulomata occur in the epididymis when the ducts rupture and release their contents into the interstitium; the mass of sperm are enclosed by macrophages, granulation tissue and inflammatory cells. A maximum incidence of 2 per cent has been seen in the AP rat. Age-related changes in the duct epithelium are common and include vacuolation and microcystic degeneration, and atrophy of the testes is reflected in the epididymides by an increasing loss of mature sperm from the

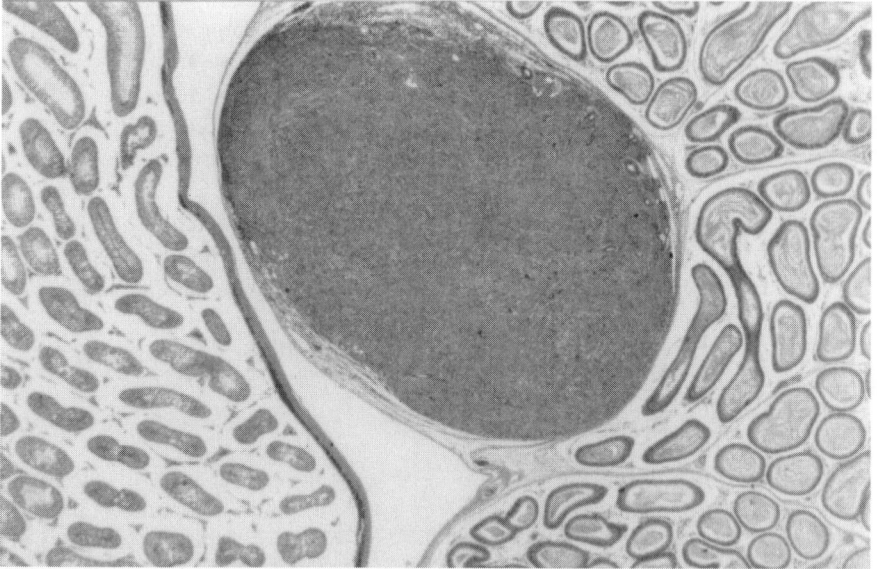

Figure 88   Fibroma of the epididymis of an AP rat. ×8, H&E

ducts. Desquamated cells from the seminiferous epithelium, cellular debris and fluid may also accumulate in the ducts. Other miscellaneous, uncommon findings include arteritis and necrosis of the epididymal fat pad.

10.2.2   *Neoplastic Changes*

Tumours of the epididymis are exceedingly rare in all strains (Mostofi and Bresler, 1976) and in the AP rat a single fibroma has been observed (Figure 88).

10.3   Seminal Vesicle

10.3.1   *Non-neoplastic Changes*

The seminal vesicles of the rat are large paired organs extending cranially from the urethral junction, and there are few reports of spontaneous lesions. A longitudinal section of the left vesicle is taken for histological examination in the AP rat. Inflammation of the seminal vesicle is uncommon (<10 per cent) in AP rats. Acute vesiculitis is less common, and when present has been secondary to infections of the urinary tract. A mild diffuse chronic vesiculitis is more common and the cause for this type is not known. This low incidence is in agreement with those published for several strains and cited by Bosland (1992).

Atrophy of the seminal vesicles is chiefly associated with androgen deprivation in the AP rat. The macroscopic appearance is of shrunken, small vesicles related to a histological appearance characterised by little secretory material within the lumen of the vesicle, a loss of epithelium and an increase in the fibromuscular stroma (Figure 89). The incidence in 2 year studies ranges between 5 and 17 per cent. Another type of atrophic change which is seen less frequently (maximum 5 per cent), is of dilated vesicles with increased secretion in the lumen and a markedly flattened secretory epithelium. Hyperplasia of the epithelium is usually focal and shows cells which are slightly disarranged from the normal glandular pattern with cellular atypia. The incidence does not exceed 5 per cent at 2 years.

10.3.2 Neoplastic Changes

Tumours of the seminal vesicle are very rare. In the AP rat two adenocarcinomas have been identified. Both showed some glandular patterns but also clumps of undifferentiated cells within an abundant fibrous stroma. Both tumours showed local extension through the capsule and into the coagulating glands, but metastases were not identified. Low incidences of seminal vesicle tumours in Wistar strains and the F344 have been cited by Bosland (1992).

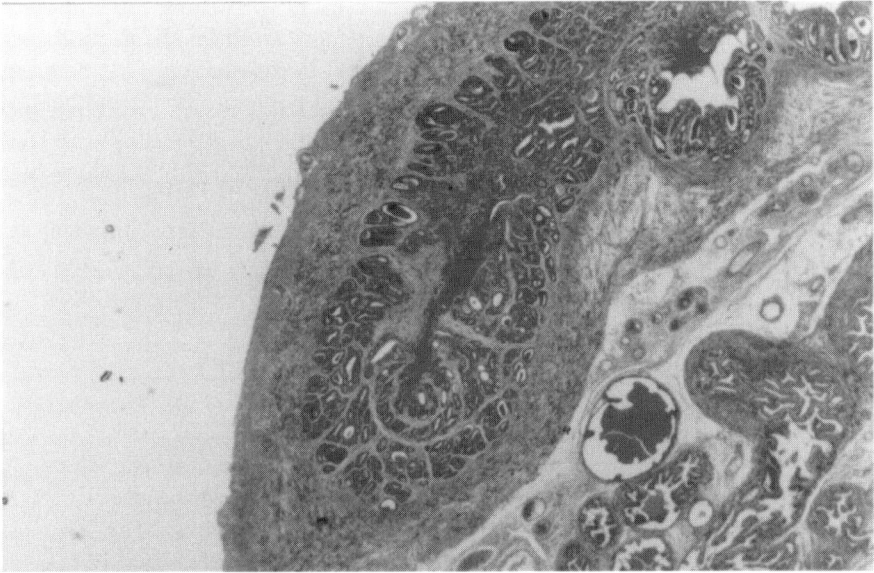

Figure 89 Atrophy of the seminal vesicle in a 26 month old AP rat: there is a marked reduction in secretory material and an increase in the fibromuscular stroma. ×8, H&E

10.4   Prostate

10.4.1   Non-neoplastic Changes

The prostate gland of the rat has four lobes: the anterior prostate or coagulating gland, the dorsal and lateral lobes, which are difficult to separate, and the ventral lobe which is the one taken for histological examination in the AP rat.

Prostate weights

The weights of the ventral prostate in the AP rat at different time points are shown in Table 10.4. There is a significant increase in the absolute weight of the ventral lobe up to 12 months and a decline in the relative weight at 12 months. As in most organs this decline in the relative weight can be related to the increase in non-lean body weight in the aging rat. This pattern of growth has been reported in COP and ACI/SegHap rats (Issacs, 1984).

Inflammation

Marked strain differences have been reported for the incidence of non-neoplastic diseases in the prostate (Burek, 1978; Issacs, 1984), and Roe (1991) noted that inflammation of the prostate is more common in obese animals. In the AP all mononuclear cell infiltrates in the interstitial tissue are not uncommon at any age (up to 10 per cent), but significant degrees of acute prostatitis are rare in young animals. In studies up to 6 months the incidence of any inflammatory lesion is below 5 per cent, but by two years it may reach levels up to 20 per cent. The inflammation is usually mild and focal in distribution. One or more acini are dilated and filled with cellular debris and polymorphs; the epithelium of affected acini is usually normal as is the surrounding interstitium. Severe acute prostatitis frequently involves the whole ventral lobe and may extend into the dorsolateral lobes; it is uncommon at any age, and in the AP rat has always been secondary to acute urinary tract infections. Spontaneous prostatitis may be associated with bacterial infection, but this is not always the cause, and low

Table 10.4   Weights of the ventral prostate in the AP rat

Age (weeks)	Number	Ventral prostate weights	
		Absolute weight (g)	Relative weight (% body weight)
12	10	0.35	0.097
34	20	0.59	0.098
58	25	0.73	0.089

temperature stress, limited space and restricted access to food and water have been shown to produce a non-bacterial inflammation in the ventral lobe in particular (Aronsonn *et al.*, 1988). The mechanism for this effect is not known, but variations in the environmental conditions in animal houses may account for the differences reported for the incidence of prostatitis in various strains, and in different studies in the same strain (Coleman *et al.*, 1977; Goodman *et al.*, 1979; Kroes *et al.*, 1981; Anver *et al.*, 1982).

Atrophy

Bosland (1992) defines three types of atrophy. In the AP rat the most common type is diffuse atrophy associated with dilatation of the acini and a flattening of the secretory epithelium, but little change in the amount of secretory material or stroma (Figure 90). It is thought that this type of change is due to a decreased rate of emptying resulting in reduced secretory activity, and it is most common in the ventral lobe (Issacs, 1984). Animals which have been castrated or have testicular atrophy show a diffuse prostatic atrophy related to androgen withdrawal. There is hyposecretion, decreased glandular size and an increase in the fibromuscular stroma, resulting in marked reduction in prostatic weight. The third type of atrophy, defined by Bosland, is a focal atrophy, similar in appearance to that which occurs in castration and is probably related to a previous focal inflammation. Dietary factors have also been shown to affect prostatic function. Long-Evans rats fed a protein-free diet for 20 days

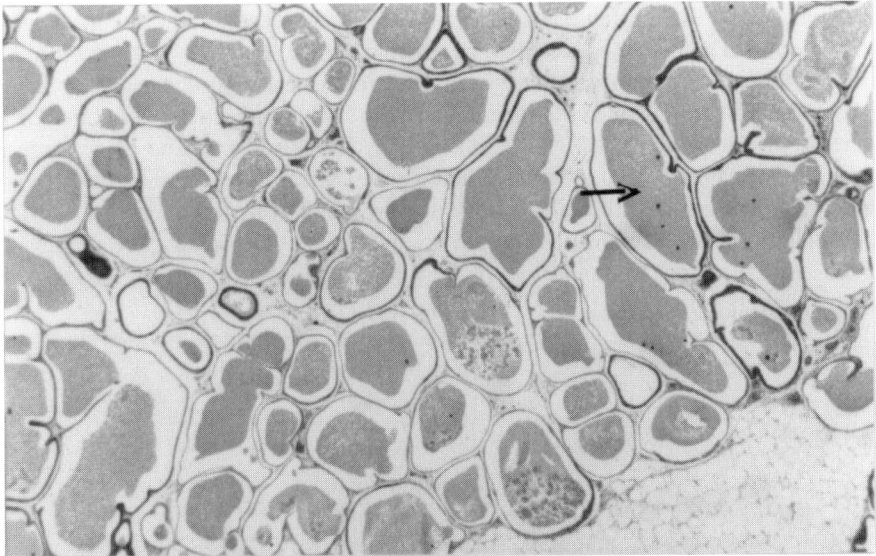

Figure 90 Diffuse prostatic atrophy in a 20 month old AP rat: acini are dilated and lined by a flattened epithelium but there is little change in the secretory material or stroma. Corpora amylacea (→) are present in several acini. ×32, H&E

show marked reductions in prostate weight (Esashi *et al*, 1982) and this also has been reported after restriction of food intake (Howland, 1975). The incidence of prostatic atrophy in the AP rat is low before 12 months but in 2 year studies reaches levels between 5 and 15 per cent. The spontaneous incidence in most strains is also low.

Corpora amylacea

Corpora amylacea are found in the prostatic acini of old AP rats (Figure 90) and some become mineralised. The numbers present vary markedly but the incidence may reach 20 per cent in 2 year studies. Bosland (1992) reported that corpora amylacea in the dorsolateral lobe may be large enough to block excretory ducts and cause inflammation and squamous metaplasia. The pathogenesis of these concretions is not known but it is age related in most strains. Issacs (1984) recorded increasing numbers in the ventral lobe with age, but decreasing numbers in the dorsolateral lobe after 25 months.

Hyperplasia

Bosland (1992) defines hyperplasia as reactive, functional or atypical. The reactive type, seen in inflammatory lesions, shows a simple increase in the cellular layers in the affected acini. The cells are usually of normal appearance and rarely show atypia. The incidence of this type is similar to the incidence of prostatitis. Functional hyperplasia produces acini with a columnar, infolded epithelium of basophilic cells. It is a rare spontaneous change but is not infrequent as a result of various treatments which interfere with hormonal homeostasis. The most common type in the AP rat is the atypical hyperplasia. There is considered to be a progression from hyperplasia to adenoma and carcinoma (Ward *et al.*, 1980; Reznik *et al.*, 1981), but as yet no clear criteria for diagnosis have been generally accepted. The distinction between hyperplasia and adenoma in the AP rat has been that the hyperplastic lesion does not completely obliterate the lumen of the acinus, does not involve more than four acini and shows no compression (Figure 91). The presence of a capsule and atypical solid growth are typical of adenomas.

Atypical hyperplasia is regarded as a pre-neoplastic state by Bosland and increases with age. In AP rats the incidence in 2 year studies is between 1 and 4 per cent and is low in most strains except the ACI/SegHap strain where the incidence is 100 per cent at 2 years (Issacs, 1984; Ward *et al.*, 1980).

10.4.2 *Neoplastic Changes*

Only a single adenocarcinoma of the ventral prostate has been observed in the AP rat. It was a small tumour, well differentiated into glandular patterns, and

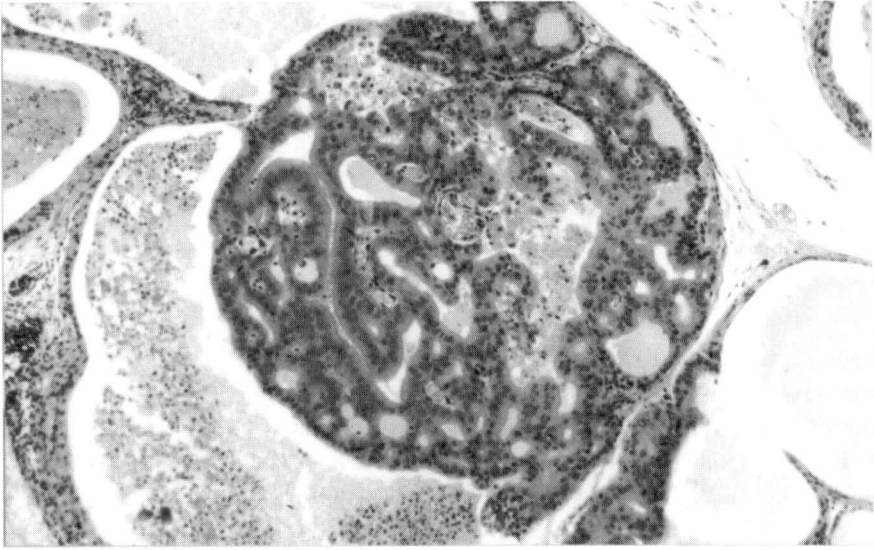

Figure 91 Hyperplasia of the prostate: the hyperplastic area does not obliterate the acini and shows no compression. ×80, H&E

showed invasion of adjacent tissue and a vein. This low incidence is true of most strains (Coleman, *et al.*, 1977; Goodman *et al.*, 1979; Kroes *et al.*, 1981; Anver *et al.*, 1982; Solleveld *et al.*, 1984).

10.5 Coagulating Gland

The coagulating gland is the anterior lobe of the prostate and is not routinely examined in the AP rat, but as it lies close to the seminal vesicle it is frequently (30 per cent) present in sections of that organ. Inflammation is an uncommon finding, usually diffuse and secondary to inflammation in the adjacent seminal vesicles. The incidence in 2 year studies is less than 1 per cent. Diffuse atrophy is seen occasionally in animals with total testicular atrophy, but no hyperplastic or neoplastic changes have been observed.

10.6 Preputial Glands

These glands are only examined when macroscopically abnormal, so it is not possible to give an accurate incidence for any lesion. Inflammation has been the most common finding, usually a diffuse acute inflammation with abscess formation. Marked duct dilatation with or without inflammation also occurs infrequently. One tumour has been observed in the preputial glands.

10.7 Penis

Very few changes have been observed in the penis, consisting of inflammatory lesions with scab formation. The macroscopic observation of changes in the penis is less than 1 per cent. No tumours have been observed.

10.8 References

AMANN, R. P. (1982) Use of animal models for detecting specific alterations in reproduction, *Fundamental and Applied Toxicology*, **3**, 13–26.

AMANN, R. P. (1986) Detection of alteration in testicular and epididymal function in laboratory animals, *Environmental Health Perspectives*, **70**, 149–58.

ANVER, M. R., COHEN, B. J., LATTUADA, C. P. and FOSTER, S. J. (1982) Age-associated lesions in barrier-reared male Sprague-Dawley rats: a comparison between Hap:(SD) and Crl:COBS®CD®(SD) stocks, *Experimental Aging Research*, **8**, 3–24.

ARONSONN, A., DAHLGREN, S., GATENBECK, L. and STROMBERG, L. (1988) Predictive sites of inflammation in the prostatic gland: an experimental study on nonbacterial prostatitis in the rat, *Prostate*, **13**, 17–25.

BARTKE, A., SWEENY, C. A., JOHNSON, L., CASTRACANE, V. D. and DOHERTY, P. C. (1985) Hyperprolactinaemia inhibits development of Leydig cell tumors in aging Fischer rats, *Experimental Aging Research*, **11**, 123–8.

BARTLETT, J. M. S., WEINBAUER, G. F. and NIESCLAG, E. (1989) Differential effects of FSH and testosterone on the maintenance of spermatogenesis in the adult hypophysectomized rat, *Journal of Endocrinology*, **121**, 49–58.

BETHEA, C. L. and WALKER, R. F. (1979) Age-related changes in reproductive hormones and in Leydig cell responsivity in the male Fischer 344 rat, *Journal of Gerontology*, **34**, 21–7.

BOMHARD, E., KARBE, E. and LOESER, E. (1986) Spontaneous tumours of 2000 Wistar TNO/W, 70 rats in two-year carcinogenicity studies, *Journal of Environmental Pathology, Toxicology and Oncology*, **7**, 35–52.

BOORMAN, G. A., HAMLIN, M. H. and EUSTIS, S. L. (1987) Focal interstitial cell hyperplasia, testis, rat, in JONES, T. C., MOHR, U. and HUNT, R. D. (Eds), *Genital System*, pp. 200–4, Berlin and Heidelberg: Springer-Verlag.

BOSLAND, M. C. (1992) Lesions in the male accessory sex glands and penis, in MOHR, U., DUNGWORTH, D. L. and CAPEN, C. C. (Eds), *Pathobiology of the Aging Rat*, Vol. 1, pp. 443–67, Washington: ILSI Press.

BUREK, J. D. (1978) Male reproductive system, in *Pathology of Aging Rats*, p. 129, Palm Beach, Florida: CRC Press.

CHAN, S. W. C., LEATHEM, J. H. and ESASHI, T. (1977) Testicular metabolism and serum testosterone in aging male rats, *Endocrinology*, **101**, 128–33.

CLEGG, E. J. (1960) The age at which male rats become fertile, *Journal of Reproduction and Fertility*, **1**, 119–20.

COLEMAN, G. L., BARTOLD, S. W., OSBALDISTON, G. W., FOSTER, S. J. and JONAS, A. M. (1977) Pathological changes during aging in barrier-reared Fischer 344 male rats, *Journal of Gerontology*, **32**, 258–78.

DYM, M., RAJ, H. F. M. and CHEMES, H. E. (1977) Response of the testis to selective withdrawal of LH or FSH using antigonadotropic sera. In TROEN, P. and NANKIN, H. R. (Eds), *The Testis in Normal and Infertile Men*, pp. 97–124, New York: Raven.

ESASHI, T., SUZUE, R. and LEATHEM, J. H. (1982) Influence of dietary protein depletion and repletion on sex organ weight of male rats in relation to age, *Journal of Nutrition Science and Vitaminology*, **28**, 163–72.

GOODMAN, D. G., WARD, J. M., SQUIRE, R. A., CHU, K. C. and LINHART, M. S. (1979) Neoplasms and non-neoplastic lesions in aging F344 rats, *Toxicology and Applied Pharmacology*, 48, 237–48.

GOODMAN, D. G., WARD, J. M., SQUIRE, P. A., CHU, K. C. and LINHART, M. S. (1980) Neoplasms and non-neoplastic changes in aging Osborne-Mendel rats, *Toxicology and Applied Pharmacology*, **55**, 433–47.

GRAY, G. D., SMITH, E. R. and DAVIDSON, J. M. (1980) Gonadotrophin regulation in middle-aged rats, *Endocrinology*, **107**, 2021–6.

GREAVES, P. and FACCINI, J. M. (1984) *Rat Histopathology*, Amsterdam: Elsevier.

HEYWOOD, R. and JAMES, R. W. (1985) Current approaches for assessing male reproductive toxicity: testicular toxicity in laboratory animals, in DIXON, R. L. (Ed.), *Reproductive Toxicology*, pp. 147–60, New York: Raven.

HOWLAND, B. (1975) The influence of feed restriction and subsequent re-feeding on gonadotrophin secretion and serum testosterone levels in male rats, *Journal of Reproduction and Fertility*, **44**, 429–36.

IMAI, K., and YOSHIMURA, S. (1988) Spontaneous tumors in Sprague-Dawley (CD:Crj) rats, *Journal of Toxicologic Pathology*, **1**, 7–12.

ISSACS, J. T. (1984) The aging ACI/Seg versus Copenhagen male rat as a model for the study of prostate carcinogenesis, *Cancer Research*, **44**, 5785–96.

JACOBS, B. B. and HUSEBY, R. A. (1967) Neoplasms occurring in aged Fischer rats, with special reference to testicular, uterine and thyroid tumours, *Journal of the National Cancer Institute*, **39**, 303–9.

JAMES, R. W. and HEYWOOD, R. (1979) Age related variations in the testes of Sprague-Dawley rats, *Toxicology Letters*,4, 257–61.

KALER, L. W. and NEAVES, W. B. (1981) The steroidogenic capacity of the aging rat testis, *Journal of Gerontology*, **108**, 712–19.

KERR, J. B., DONACHIE, K. I. and ROMMERTS, F. F. G. (1985) Selective destruction and regeneration of Leydig cells *in vivo*. A new method for the study of seminiferous tubular-interstitial tissue interaction, *Cell and Tissue Research*, **242**, 145–56.

KROES, R., GARBIS-BERKVENS, J. M., DE VRIES, T. and VAN NESSELROOIJ, J. H. J. (1981) Histopathological profile of a Wistar rat stock including a survey of the literature, *Journal of Gerontology*, **35**, 259–79.

LAMB, J. C. and CHAPIN, R. E. (1985) Experimental models of male reproductive toxicology, in THOMAS, J. A., KORACH, K. S. and MCLACHLAN, J. A. (Eds), *Endocrine Toxicology*, pp. 85–115, New York: Raven Press.

LEBLOND, C. P. and CLERMONT, C. P. (1952) Definition of stages of the cycle of the seminiferous epithelium in the rat, *Annals of the New York Academy of Sciences*, **55**, 548–72.

MAEKAWA, A., KUROKAWA, Y., TAKAHASHI, M., KUKOBO, T., OGIU, T., ONODERA, H., TANIGAWA, H., OHNO, Y., FURUKAWA, F. and HAYASHI, Y. (1983) Spontaneous tumours in F344/DuCrj rats, *Gann*, **74**, 365–72.

179

MAITA, K., HIRANO, M., HARADA, T., MITSUMORI, K., YOSHIDA, A., TAKAHASHI, K., NAKASHIMA, N., KITAZAWA, A., INUI, K. and SHIRASU, Y. (1987) Spontaneous tumours in F344/DuCrj rats from 12 control groups of chronic and oncogenicity studies, *The Journal of Toxicological Science*, **12**, 111–26.

MILLS, N. C., MILLS, T. M. and MEANS, A. R. (1977) Morphological and biochemical changes which occur during post natal development and maturation of the rat testis, *Biology of Reproduction*, **17**, 124–30.

MOSTOFI, F. K. and BRESLER, V. M. (1976) Tumours of the testis, in TURUSOV, V. S. (Ed.), *Pathology of Tumours in Laboratory Animals*, Vol. 1, *Tumours of the Rat*, pp. 135–60, Lyon: IARC.

REZNIK, G. K., HAMLIN, M. H., WARD, J. M. and STINSON, S. F. (1981) Prostatic hyperplasia and neoplasia in aging F344 rats, *The Prostate*, **2**, 261–8.

ROBB, G. W., AMANN, R. P. and KILLIAN, G. J. (1978) Daily sperm production and epidiymal sperm reserves of pubertal and adult rats, *Journal of Reproduction and Fertility*, **54**, 103–7.

ROE, F. J. C. (1991) *1200 Rat-Biosure Study, Design and Overview of Results in Biological Effects of Dietary Restriction*, pp. 287–304, Berlin and Heidelberg: Springer-Verlag.

RUSSELL, L. D. (1994) Normal development of the testis, in MOHR, U., DUNGWORTH, D. L. and CAPEN, C. C. (Eds), *Pathobiology of the Aging Rat*, p. 399, Washington: ILSI Press.

RUSSELL, L. D., MALONE, J. P. and KARPAS, S. L. (1981) Morphological pattern elicited by agents affecting spermatogenesis by disruption of its hormonal stimulation, *Tissue and Cell*, **13**, 369–80.

SASS, B., RABSTEIN, L. S., MADISON, R., NIMS, R. M., PETERS, R. L. and KELLOFF, G. J. (1975) Incidence of spontaneous neoplasms in F344 rats throughout the natural life-span, *Journal of the National Cancer Institute*, **54**, 1449–56.

SHARPE, R. M. (1989) Follicular stimulating hormone and spermatogenesis in the adult male, *Journal of Endocrinology*, **121**, 405–7.

SHARPE, R. M., DONACHIE, K. and COOPER, I. (1988) Re-evaluation of the intratesticular levels of testosterone required for maintenance of spermatogenesis in the rat, *Journal of Endocrinology*, **117**, 19–26.

SOLLEVELD, H. A., HASEMAN, J. K. and MCCONNELL, E. E. (1984) Natural history of body weight gain, survival and neoplasia in the F344 rat, *Journal of the National Cancer Institute*, **72**, 929–40.

THURMAN, J. D., MOELLER, R. B. JR. and TURTURRO, A. (1995) Proliferative lesions of the testis in ad libitum-fed and food restricted Fischer 344 and FBNF1 rats, *Laboratory Animal Science*, **45**(6), 635–46.

TUCKER, M. J. and ORTON, T. C. (1995) *Comparative Toxicology of Hypolipidaemic Fibrates*, pp. 47–53, London: Taylor & Francis.

TUREK, F. W. and DESJARDINS, C. (1979) Development of Leydig cell tumors and onset of changes in the reproductive and endocrine systems of aging F344 rats, *Journal of the National Cancer Institute*, **63**, 969–75.

WARD, J. M., REZNIK, G., STINSON, S. F., LATTIRADA, C. P., LONGFELLOW, D. G. and CAMERON, T. P. (1980) Histogenesis and morphology of naturally occurring prostatic carcinoma in the ACI/segHapBR rat, *Laboratory Investigation*, **43**, 517–22.

WRIGHT, J. R., YATES, A. J., SHARMA, H. M., SHIM, C., TIGNER, R. L. and THIBERT, P. (1982) Testicular atrophy in the spontaneously diabetic BB Wistar rat, *American Journal of Pathology*, **108**, 72–9.

ZENICK, H. and CLEGG, E. D. (1989) Assessment of male reproductive toxicity: a risk assessment approach. In HAYES, A. W. (Ed.), *Principles and Methods of Toxicology*, 2nd edition, pp. 275–309, New York: Raven.

11

The Endocrine System

The endocrine system of the rat, as in other species, is a complex series of inter-relationships between the pancreatic islets, adrenal, thyroid and parathyroid glands, and the pituitary gland. The pituitary is controlled, via the hypothalamus, by input from various areas of the brain, not all of which are fully characterised. Endocrine glands are exquisitely sensitive to changes in the stimulating or inhibiting hormones which control their function, and spontaneous non-neoplastic changes and tumours are an important feature of the morbidity and mortality in the rat.

11.1 Pituitary Gland

11.1.1 Non-neoplastic Changes

The pituitary gland is the most complex of the endocrine glands in terms of structure and function, and in older rats hyperplasia and adenomas are common in all strains. This is in contrast to the human pituitary gland where there is a mild atrophy of the anterior lobe and a reduction in the numbers of tumours in old age.

Pituitary weights

The weights of the pituitary gland at various time points up to 1 year are show in Table 11.1. The absolute and relative weights of the pituitary gland female are greater than those of the male at all times. The maximum we' the gland occurs at 6 months; thereafter the relative weight declines as '

Table 11.1 Weights of the pituitary gland in the AP rat

Age (weeks)	Number	Weight of pituitary			
		Absolute weight (g)		Relative weight (% body weight × 1000)	
		Male	Female	Male	Female
12	10	0.010	0.010	2.7	4.8
34	20	0.012	0.016	1.9	5.2
58	25	0.014	0.015	3.6	4.9

gain non-lean body weight. The pituitary gland is not weighed in 2 year studies as the presence of tumours obscures any underlying changes in the gland.

Structure and functional changes

The pituitary gland of the rat develops from Rathké's pouch and is composed of two lobes: the anterior lobe (adenohypophysis) and the posterior lobe (neurohypophysis) which lies in a horizontal position to the hypothalamus. The anterior lobe is further divided into the anterior part (pars distalis) and an intermediate lobe (pars intermedia). The anterior lobe has cells which secrete at least six different peptide hormones: growth hormone or somatotrophin (GH), prolactin (PRL), adrenocorticotrophic hormone (ACTH), thyroid stimulating hormone (TSH), follicular stimulating hormone (FSH) and luteinising hormone (LH). In sections stained with H&E the acidophilic cells are GH and PRL secreting cells and the basophils are the cells which secrete the other four hormones. The cells which contained no stain (chromophobe cells) were once thought to be inactive hormonally, but capable of transforming into other secreting cells. The chief cell of the intermediate lobe mainly secretes melanocyte stimulating hormone (MSH) and β-endorphin. The advent of immunocytochemistry, using antisera to the hormones of the anterior lobe, is an important means of accurate characterisation, and immunoreactivity identifies five different cell types. These cells are the somatotrophs (GH secreting), lactotrophs (PRL secreting), corticotrophs (ACTH secreting), thyrotrophs (TSH secreting) and the gonadotrophs which produce FSH and LH. In the male rat somatotrophs are the predominant cell in the anterior lobe (50 per cent in males and 40 per cent in females). In the female, lactotrophs are the most numerous cells (40 per cent). In the male the majority of gonadotrophs are bihormonal, producing both FSH and LH; in the female approximately 25 per cent produce either FSH or LH (Childs *et al.*, 1980). The morphology, ultrastructure and immunocytochemisty have been described in detail by Stefaneanu and Kovacs (1994). The neurohypophysis produces two hormones: anti-diuretic hormone (ADH or vasopressin) and oxytocin. The complex regulation of pituitary hormone secretion is not within the scope of this volume, but it has been described by Lechan (1987) and Reichlin (1989).

Congenital abnormalities

Cysts are the only congenital abnormalities which have been observed in the pituitary of the AP rat. These include Rathké's pouch remnants which have a pseudostratified ciliated columnar epithelium, and craniopharyngeal ducts which have a ciliated epithelium and contain mucin. They are usually small, generally only detected at microscopic examination, and are unlikely to have any serious effects on the health of the rat unless they are markedly dilated, when they may cause compression atrophy of the anterior lobe or haemorrhage. In the AP rat the maximum incidence in a 2 year study is 5 per cent. They have been reported in other strains (Carlton and Gries, 1983).

Degenerative changes

Inflammation and necrosis are very rare (<1 per cent) in the pituitary gland of the AP rat, but have been observed in occasional animals. In all cases there was no obvious cause for either of these conditions. It is known that the lymphocytic choriomeningitis virus replicates in the somatotrophs of the anterior pituitary, but there is no associated cell necrosis or inflammation (Oldstone *et al.*, 1982).

Cellular changes

Castration cells have been observed in the pituitary glands of AP rats which have severe testicular atrophy. These cells are enlarged with a characteristic 'signet ring' appearance (Figure 92). These cells have been extensively investigated in studies of castrated rats, and immunohistochemical staining indicates that they are gonadotrophs which contain decreased numbers of secretory granules and dilated rough endoplasmic reticulum (Ibrahim *et al.*, 1986). Rats which have been exposed to low barometric pressure have shown a 30 per cent reduction in thyrotrophs which is not associated with any important cytological changes (Gosney, 1986). The pituitary gland is highly susceptible to the effects of xenobiotics on the different cell populations. Oestrogens increase the numbers of prolactin cells (Lloyd, 1983) although there are strain differences in response (Stone *et al.*, 1979; Wiklund *et al.*, 1981). Administration of goitrogens or thyroidectomy increases the numbers of thyrotrophic cells (Purves and Griesbach, 1956; Lloyd and Mailloux, 1987), and aminoglutethamide increases the corticotrophs (Zak *et al.*, 1985).

Hyperplasia

Hyperplasia of the pituitary gland is generally considered to be an intermediate stage in the development of tumours. As in all similar neoplastic progressions

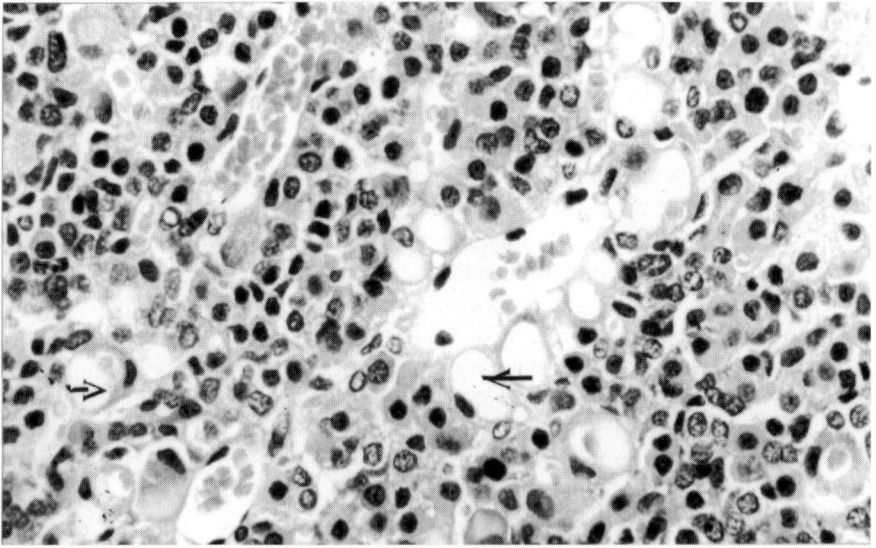

Figure 92   Castration cells (←) with the 'signet-ring' appearance in the pituitary gland of a male AP rat with testicular atrophy. ×128, H&E

the distinction between hyperplasia and microadenoma of the pituitary gland is confined to the size and the presence of compression of adjacent cells by the tumour.

11.1.2   *Neoplastic Changes*

The incidence of tumours of the pituitary gland is high in all strains of rat (Furth *et al.*, 1976), and in the AP rat tumours of the pars distalis are the most common type of tumour in the strain, and the largest single cause of mortality. The clinical signs associated with the presence of large pituitary adenomas are body weight loss (pituitary cachexia), muscle atrophy, red staining around the eyes and a variety of neurological signs. Mortality may be due to haemorrhage into the brain, but the more common cause is the increased intercranial pressure which produces a general deterioration in health. The tumours range in size from microadenomas which are only detected by microscopic examination, through small nodules (Figure 93), to very large tumours which cause severe cerebral compression. The tumours are almost always benign. Histologically there appear to be two distinct types. The most common type is the haemorrhagic adenoma which has cords of cells around large blood-filled spaces (Figure 94) and the less common solid tumour composed of sheets of cells showing no structure and no blood-filled spaces (Figure 95). In H&E stained sections the tumours appear to be chromophobic, but immunostaining of approximately one hundred adenomas from untreated AP rats has shown around 60 per cent of the

186

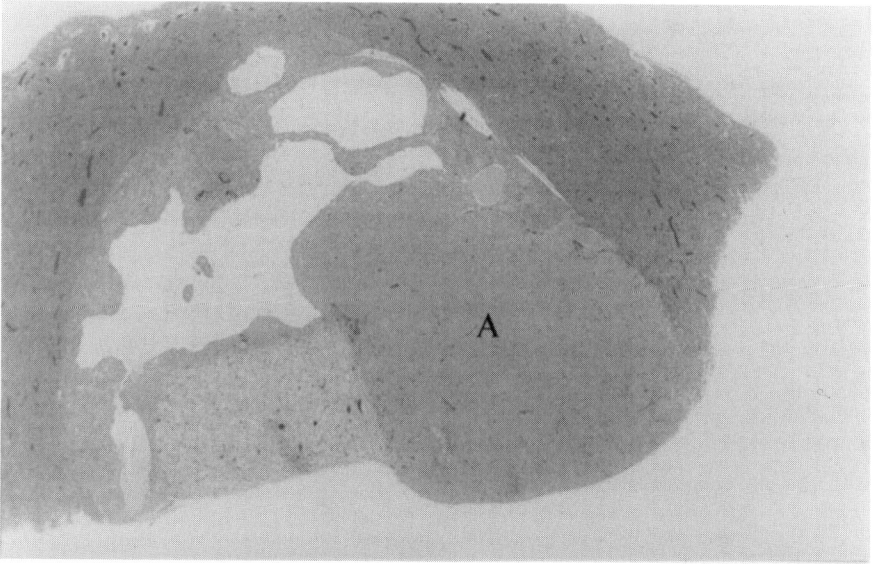

Figure 93   Microadenoma (A) of the pituitary in an AP rat aged 16 months. ×32, H&E

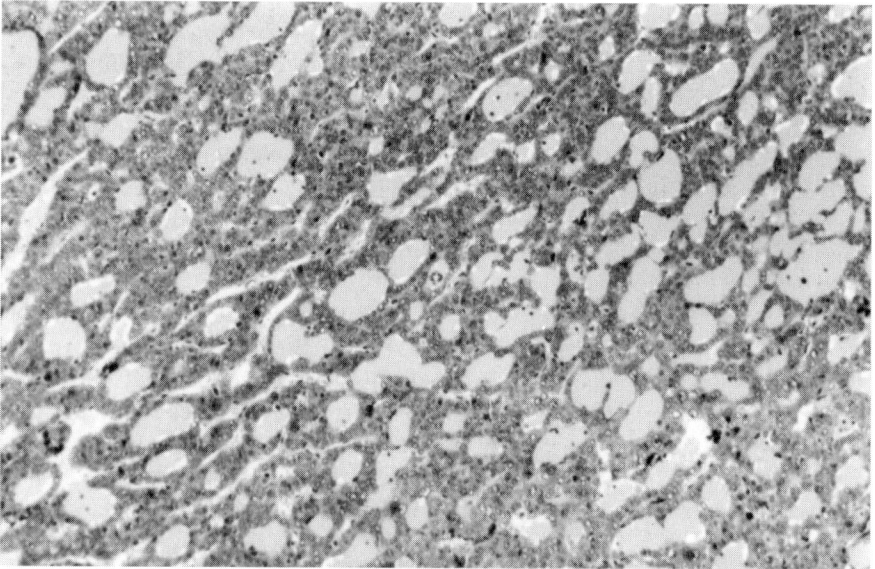

Figure 94   Haemorrhagic adenoma of the pituitary gland showing cords of cells around spaces which may be filled with blood. ×80, H&E

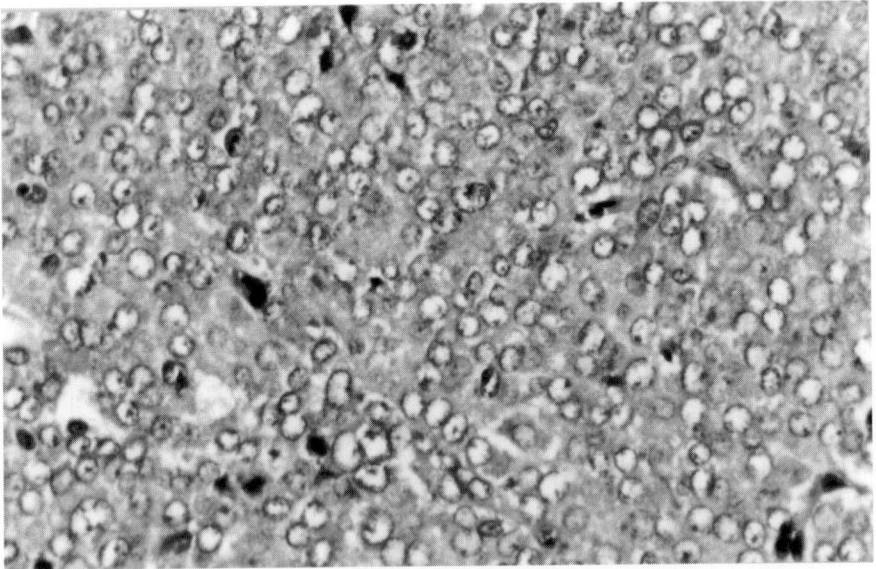

Figure 95 Solid pituitary adenoma of compact masses of uniform cells. ×120, H&E

tumours were positive for PRL, the remaining 40 per cent were immuno-negative (30 per cent) or positive for TSH or LH (10 per cent). These results are in good agreement with several more comprehensive published studies of pituitary tumour immunostaining (Berkvens *et al.*, 1980; McComb *et al.*, 1984; Sandusky *et al.*, 1988) which demonstrate that the majority are of PRL-secreting cells. An exception to this are male Lobund-Wistar rats. In this strain, pituitary tumours in males are most frequently of gonadotrophs (Stefaneanu and Kovacs, 1994). Metastases from pituitary adenomas have not been observed, but a few (<2 per cent) have been classified as carcinomas on the basis of infiltration into the brain. A similar low incidence of carcinomas was reported in SD rats by Magnusson *et al.* (1979).

The time of onset of hyperplasia in the AP rat is approximately 6 months and the first tumours are also seen at this time, but they remain few in number until 18 months of age. Tumours, when small, may be multifocal, but these have been classified as one tumour. The incidence is always higher in females and there has been a marked increase in incidence over the three decades the AP rat has been used in toxicology studies. This is illustrated in Figure 96, which shows the percentage incidence of pituitary adenomas in the control animals of studies completed in the designated years (there were 100 or 150 animals at each time point).

In the first carcinogenicity studies in the AP rat in 1960 the incidence was approximately 10 per cent in males and 20 per cent in females. By 1980 the incidence had risen to nearly 60 per cent in males and over 90 per cent in

Incidence of Spontaneous Pituitary Tumours in the AP Rat

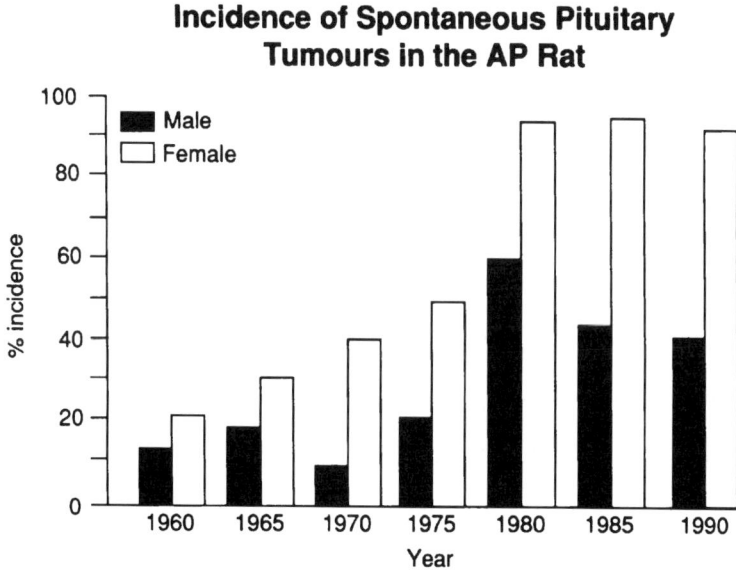

Figure 96 Histogram of the percentage incidence of pituitary adenomas in control AP rats in studies completed between 1960 and 1990

females. The incidence has changed little between 1980 and 1990. A similar trend can be seen in the published data for outbred strains such as the Wistar and SD, while incidence levels in the inbred F344 strain are lower and appear little changed with time (Table 11.2).

Table 11.3 shows the incidence of tumours in the life-span study completed in the early 1960s. The first pituitary tumour was seen at 18 months and more than 60 per cent occurred in animals over 30 months old (Table 11.3). There is also a marked difference in the incidence in breeding and non-breeding females, with non-breeding females having more than twice the incidence found in the breeding animals. The reason for this protective effect of pregnancy is not known but it has also been observed by Pickering and Pickering (1984).

The pathogenesis of pituitary adenomas is not known. It is suggested that as they are mostly PRL secreting it is likely to be due to a reduction in the hypothalamic secretion of dopamine, since this is the hormone which inhibits prolactin (Prysor-Jones *et al.*, 1983). As the tumours are more common in females, oestrogens are likely to be involved. Administration of oestrogens has been reported to produce pituitary tumours (Furth *et al.*, 1956; Lloyd, 1983), although the incidence varies with the oestrogen, the dose, duration of treatment, and the strain. Some workers also consider that the tumours produced by oestrogens are not really neoplasms since they are not always autonomous and may regress on withdrawal of oestrogen (Treip, 1983; Lumb *et al.*, 1985). The mechanism is thought to be the inhibition of dopamine by

189

Table 11.2 Incidence of pituitary tumours in Fischer 344, Sprague-Dawley and Wistar rats

Strain	Male		Female		Reference
			Number of pituitary tumours		
Fischer 344	9/96	(9%)	31/102	(30%)	Jacobs and Huseby, 1967
	38/160	(23%)	69/192	(35%)	Sass *et al.*, 1975
	26/100	(26%)	36/100	(36%)	Sandusky *et al.*, 1988
Sprague-Dawley	111/345	(32%)	146/265	(55%)	Magnusson *et al.*, 1979
	74/120	(62%)	100/120	(83%)	Dodd *et al.*, 1987
	133/207	(64%)	145/207	(70%)	Greenhill, 1992
Wistar	28/300	(9%)	55/300	(18%)	Ueberberg and Lutzen, 1979
(CFB)	19/197	(10%)	37/182	(20%)	Kroes *et al.*, 1981
(SPF-Tox)	66/192	(34%)	105/192	(55%)	Kroes *et al*, 1981
	138/1020	(13%)	363/1145	(32%)	Barsoum *et al.*, 1985

Table 11.3 Incidence of pituitary tumours in a life-span study in AP rats

Age in months	Males		Females	
	Non-breeding	Breeding	Non-breeding	Breeding
0–24	0	0	4	2
25–30	17	0	29	6
31–36	13	2	33	18
37–42	4	5	8	3
43–51	0	2	1	2
Total	34/152 22%	9/42[a] 21%	75/137 55%	31/151 21%

[a] There were fewer breeding males as they were kept in harems (one male to four females).

oestrogen, yet in female rats the pituitary adenomas develop in older animals as ovarian function and oestrogen production decline. Other factors which have been shown to affect pituitary tumour incidence include dietary restriction. This has been shown to decrease the incidence of pituitary adenomas in both sexes. AP rats fed 20 per cent less than *ad libitum* controls showed a significant reduction in pituitary tumours after 2 years (Tucker, 1979), and low protein diets also reduce the incidence (Berry, 1986). Conybeare (1988) has shown that at 1 year circulating prolactin levels were elevated in *ad libitum* fed rats, compared with restricted animals, and

190

dopamine receptors were reduced. These findings all suggest that the increase in incidence of pituitary tumours in the AP rat is related to the increase in body weight which has occurred in the strain over the 30 years it has been used in toxicology studies.

Other pituitary tumours

Two adenomas of the intermediate lobe have been found in the AP rat. Both were immunoreactive for ACTH. Similar low incidences have been reported by other workers, e.g. Berkvens *et al.* (1980) found four intermediate lobe tumours in 114 Wistar rats. Two craniopharyngiomas with extensive squamous cell differentiation have been seen in the pars distalis. No tumours have been recorded in the neurohypophysis.

11.2   Pancreatic Islets of Langerhans

The islets of Langerhans comprise only 1–2 per cent of the pancreas, with the population of islets highest in the tail. Sampling of the pancreas has varied to some extent in the AP rat, but the sample is usually taken from the pancreatic tissue located in the concavity of the duodenum. Four different cell types have been located in the islets and can be identified readily by immunostaining. The majority are the insulin-secreting beta cells which constitute about 80 per cent of the islets and are located centrally. The peripherally located alpha cells produce glucagon and constitute about 15 per cent of the total. Islets in the tail have more glucagon-secreting cells than islets located in the head. The somatostatin-producing delta cells are also located at the periphery and constitute around 4 per cent of the islet. The remaining cells are pancreatic polypeptide (PP) cells which are most common in the islets in the head of the pancreas. Insulin is involved in many metabolic functions and promotes the uptake of glucose and the synthesis of glycogen, fat and protein in many tissues. Glucagon promotes the production of glucose from hepatic glycogen and other gluconeogenic precursors, and somatostatin is a potent inhibitor of insulin and glucagon secretion, indicating that it has paracrine activity in the islets. Little is known, at present, of the function of PP but it appears to have some metabolic activity and is secreted in response to protein.

11.2.1   Non-neoplastic Changes

Congenital anomalies

Ectopic pancreas has been found in the small intestine of AP rats. It is a rare condition (<1 per cent).

Inflammation

Inflammation of the islets is a very rare spontaneous condition in the AP rat and is usually focal, involving only a few islets, rather than a diffuse condition. In most cases it occurs as a secondary infiltration from an inflammatory lesion in the exocrine pancreas. This low incidence is common to most rat strains, except the BB Wistar rat which develops an insulitis involving a generalised infiltration of the islets by small lymphocytes and monocytes (Wright *et al.*, 1985). This resembles the inflammatory changes seen in juvenile diabetes in man, and the strain has been used extensively as a model of diabetes mellitus.

Foci of pigment (iron) laden macrophages are frequently seen at the edge of hyperplastic islets in older AP rats and have been reported in other strains (Hadju and Rona, 1967). They may be related to small haemorrhages caused by the expanding islet.

Metaplasia/hyperplasia

Hepatocyte metaplasia has been reported in the islets of F344 rats (Riley and Boorman, 1990). The hepatocytes appear at the periphery of the islet and may be one to four cells thick.

Hyperplasia is the most frequently seen non-neoplastic condition in the islets of aging rats, and in AP rats reaches levels up to 10 per cent in 2 year studies. Histologically the architecture of hyperplastic islets is relatively normal and may be enlarged in an even, rounded fashion, but more frequently they show a

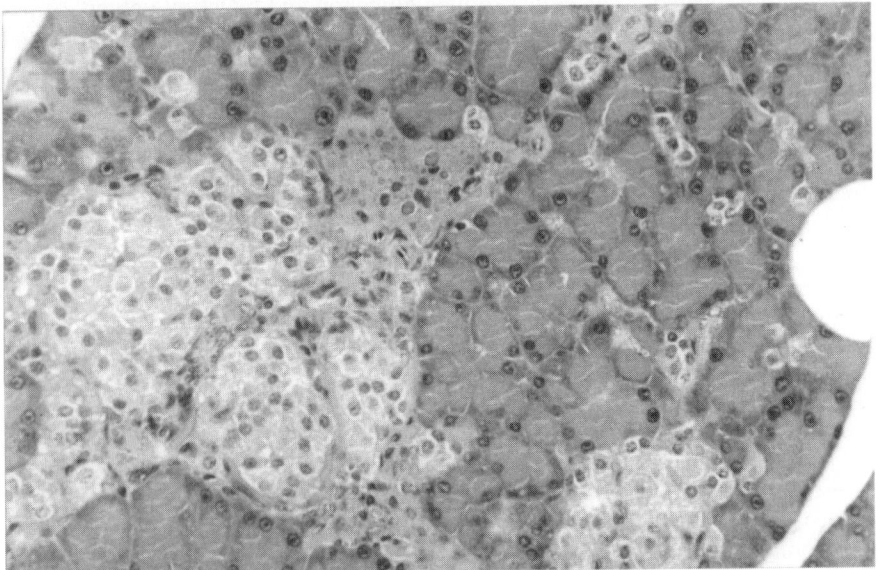

Figure 97 Lobular hyperplasia of a pancreatic islet of Langerhans in a male AP rat. ×80, H&E

mutilobular appearance (Figure 97). Immunostaining has demonstrated that this is β cell hyperplasia. It has been shown that the chief functional change in the islets of older rats is a decline in insulin secretion (Coleman *et al.*, 1977; Draznin *et al.*, 1985). This has been shown in F344 rats (Draznin *et al.*, 1985; Wang et al., 1988), SD rats (Molina *et al.*, 1985; Sartin *et al.*, 1986) and Wistar rats (Verspohl and Ammon, 1983). The cause of the hyperplasia has not yet been defined but the evidence suggests that the β cells are not defective and that the decline in insulin secretion may be due to changes in paracrine control by glucagon and somatostatin (Chaudhuri *et al.*, 1983). The decreased secretion of insulin by individual β cells may cause a compensatory hyperplasia of the cells to maintain normal levels of insulin.

11.2.2 Neoplastic Changes

Islet cell tumours in the rat are single tumours and may be distinguished from hyperplastic areas by their size and the presence of compression and capsule formation, although these are variable features. The cells of the tumour differ little from normal islet cells although they may have larger, vesicular nuclei, and mitotic activity is not prominent. The tumours have been shown to be multihormonal with β cells as the large majority, but with occasional glucagon- and somatostatin-secreting cells (Spencer *et al.*, 1986). Metastases are considered by some workers to be the only criterion for malignancy, and they consider that islet cell tumours which show small groups of cells invading the surrounding capsule or adjacent tissue are not demonstrating real malignant behaviour. Malignant tumours of the islets have been seen in two AP rats which had metastases to the liver or lung. The incidence in AP rats is always higher in males, with the incidence in 2 year studies ranging between 2 and 8 per cent in males and 0 to 2 per cent in females. In the life-span study completed in the early 1960s, the incidence in non-breeding males was 6 per cent compared with 2 per cent in breeding males, and 3 per cent in non-breeding females compared with 0.6 per cent in breeding females. This would suggest that breeding has had some influence in the development of islet cell tumours. There has been no significant change in the incidence over the three decades the AP rat has been used in toxicology studies.

The incidence in the AP rat is similar to that reported for other strains, where a preponderance was also seen in males. A 6 per cent incidence was reported for F344 rats (Stromberg *et al.*, 1983) and Wistar WAG/Rij rats (Burek, 1978) and SD and Long-Evans rats (Spencer *et al.*, 1986).

11.3 Adrenal Gland

Both adrenal glands have been examined in toxicology studies in the AP rat, with a requirement that histological sections pass through the centre of the glands to demonstrate the cortex and medulla. The highly vascularised cortex occupies two-thirds of the gland, and the medulla the remaining third. These two areas have quite distinct embryological origins and endocrine function. In the rat the cortex has three zones. The zona glomerulosa, which lies immediately below the thin fibrous capsule, constitutes approximately 15 per cent of the cortex. Below this is the largest zone, the zona fasciculata, which contributes about 70 per cent of the cortical tissue. The cells are arranged in radial cords, are larger than those of the zona glomerulosa and contain abundant lipid droplets. The inner zona reticularis makes up the remaining 15 per cent. The distinction between the zonas fasciculata and reticularis in H&E-stained sections is not clear cut. The cells tend to be smaller and there are subtle tinctorial differences. The medulla is composed chiefly of chromaffin cells and some ganglion cells, but finger-like projections or small groups of cortical cells may be present in the medulla.

11.3.1 *Non-neoplastic Changes*

Adrenal weights

The weights of the adrenal gland at various time points are shown in Table 11.4. Adrenal weights are greater in females at all times, but in both sexes they increase in absolute weight, from 12 weeks, while the relative weight decreases. This is different from the adrenal weights reported for SD rats by Yarrington and Johnston (1994) where weights only increased until 90 days and females showed a marked decrease in weight after 1 year. It has been shown that in SD rats the change in adrenal weight is primarily due to changes in the size, rather than numbers, of cortical cells (Reaven *et al.*, 1988). This hypertrophy of cells in the inner two zones is thought to be due to increasing circulating levels of ACTH (Rebuffat *et al.*, 1992). The sexual dimorphism in the adrenal gland weight of rats is thought to be due to the effects of oestrogen in the female.

Table 11.4 Weights of the adrenal glands in the AP rat

| Age (weeks) | Number/ sex | Median weight adrenal glands | | | |
| | | Weight (g) | | Relative weight (% body weight) | |
		Males	Females	Males	Females
12	10	0.062	0.078	0.017	0.035
34	20	0.064	0.098	0.011	0.030
58	25	0.066	0.087	0.008	0.021
110	50	0.071	0.093	0.012	0.019

Functional changes

The adrenal cortex secretes steroid hormones, derived from cholesterol, and these are an important component of the physiological response to stress, while the medulla primarily secretes catecholamines. The adrenal cortex can produce up to 50 different steroid hormones but the most important are aldosterone, which is a mineralocorticoid, and corticosterone, a glucocorticoid. In many species the most abundant steroid hormone is the androgen dehydroepiandrosterone sulphate (DHAS), but the rat lacks the 17α-hydroxylase activity necessary for this steroid and androgen production is minimal. No significant storage of aldosterone or corticosterone occurs in the adrenal cortex, which only stores the precursor, cholesterol. Continual synthesis of steroids is essential to maintain homeostasis. By contrast, medullary cells store catecholamines ready for rapid release, so that in the medulla synthesis and secretion are separate activities. The structure of the adrenal cortex changes in response to levels of ACTH: a marked reduction in the hormone causes atrophy due to both loss of cells and reduced cell size in the zonas fasciculata and reticularis; high levels of ACTH increase the size and number of cells in these zones. Changes in the appearance of the zona glomerulosa in ACTH deficiency are only transient. After hypophysectomy the appearance of the glomerulosa cells returns to normal within a few weeks. Electrolyte status profoundly affects the cells of the zona glomerulosa since the function of the hormone produced by this zone, aldosterone, is the control of ion transport. Sodium deficiency and potassium increase produce an increase in the width of the glomerulosa due to an increase in cell size and number. The metabolic pathways for steroid biosynthesis are complex but have been described in detail by Yarrington and Johnston (1994) and Hinson and Raven (1996).

The functions of the steroids are widespread, with important activities in carbohydrate metabolism, elevating blood sugar and promoting gluconeogenesis from protein breakdown. In addition they have anti-inflammatory and immunosuppressive functions. Although the role of the steroids in some processes is not clear, they are important in modulating stress (Munck *et al.*, 1984). The secretion of corticosterone has a circadian pattern, and in the rat, which is a nocturnal animal, the peak is in the evening and the nadir in the morning. This should be considered when investigating cortical function.

Secretion of the catecholamines adrenaline and noradrenaline, from the medulla, is regulated by the splanchnic nerve, which is stimulated by many factors (Winkler *et al.*, 1986), including stress, hypothermia and hypoglycaemia. Increased levels of catecholamines, related to increased splanchnic nerve discharges, occur after the age of 300 days (Ito *et al.*, 1986). Elevated levels have also been reported by Kvetnansky *et al.* (1971) in rats stressed by immobilisation. Strain differences have been demonstrated in the response of the rat to stress. Wistar-Kyoto rats showed two-fold higher levels of catecholamines in response to foot shock compared with Brown Norway rats (McCarty and Kopin, 1978).

Congenital anomalies

Accessory adrenocortical nodules are not uncommon in the AP rat or other strains (Dribben and Wolfe, 1947). The nodules are composed of apparently normal cortical tissue and are usually attached to the adrenal capsule or lie in the periadrenal fat. They can be confused with protruding hyperplastic nodules or neoplasms. Ectopic bone and foci of tubular structures, resembling renal tubules, have also been seen occasionally in the cortex.

Capsule

The most common change in the aging rat adrenal is thickening of the capsule by proliferation of collagen (Dribben and Wolfe, 1947). The reported incidence of this change is variable as many workers do not record it since they consider it an essentially normal feature of the aging adrenal gland.

Necrosis and inflammation

Small inflammatory cell infiltrates, usually of small lymphocytes, are seen in the cortex and medulla of up to 10 per cent of AP rats. Necrosis of the cortex is also an uncommon condition in the AP rat (maximum incidence 5 per cent) and has only been observed as a secondary result of infarction. Post-necrotic scarring with mineralisation occurs in long-standing lesions. Necrosis of the medulla has only been observed in two AP rats. It was not associated with any other obvious pathological change.

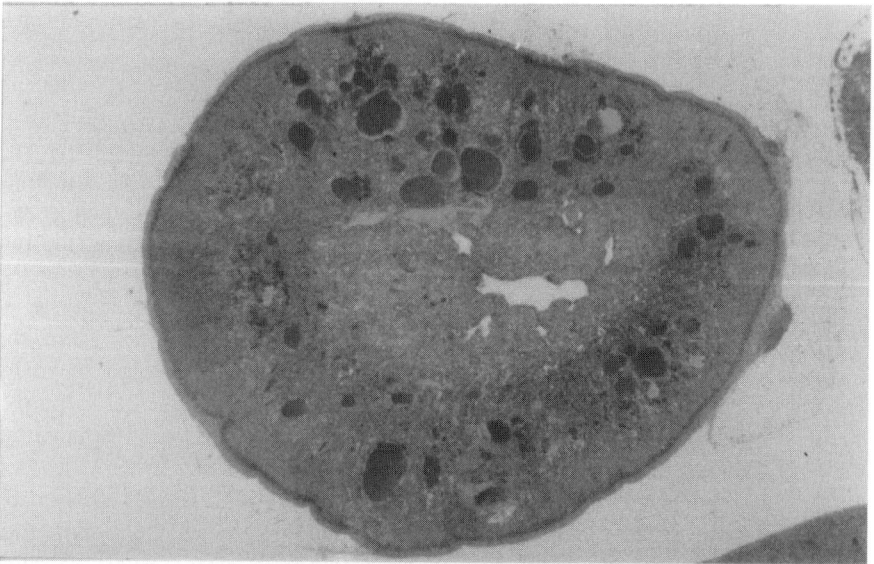

Figure 98 Multifocal cortical haemocysts in the adrenal gland of a female AP rat aged 26 months. ×8, H&E

Haemocysts (telangiectasis, peliosis)

Haemocysts are the most common histological finding in the adrenal of the AP rat. They are characterised by blood-filled spaces lined by endothelium, are of varying size, and may contain blood or thrombi (Figures 98 and 99). In the AP rat they appear after 12 months and are more common in females, with a maximum incidence of 90 per cent in a 2 year study compared with 30 per cent in males. The incidence has increased significantly in the last 30 years from levels of 10 per cent and 2 per cent in females and males, respectively, in the life-span study completed in the 1960s. Dhom *et al.* (1981) described the ultrastructure of the cysts and considered that they have features of pericapillary oedema and capillary collapse. Burek (1978) recorded a 5 per cent incidence in BN/Bi and the Wistar WAG/Rij.

Serous cysts on the capsule are seen in up to 1 per cent of AP rats. These have a thin wall, are lined by a single endothelial layer, and contain serum. They may be quite large, and visible macroscopically.

Fatty vacuolation (lipidosis)

The cells of the adrenal cortex are rich in lipid droplets, of 0.5 µm diameter, which are composed of esterified cholesterol (Nussdorfer, 1980). They increase in old age, and this may be associated with impaired steroid metabolism in older animals. Oestrogens can produce this effect in the cortical cells by inhibition of cholesterol metabolism and increased storage of non-utilised steroid precursors. The absence of the negative feedback of circulating

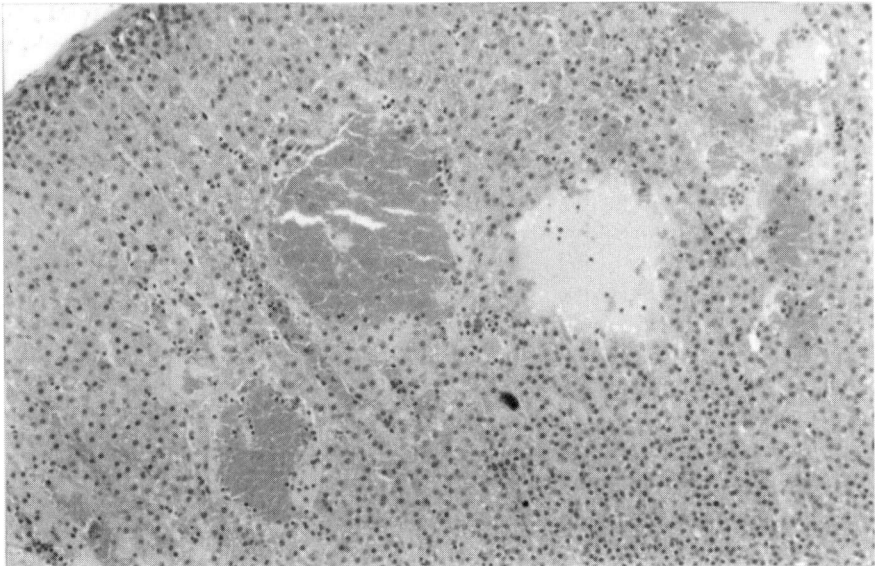

Figure 99 Higher power view of cortical haemocysts. ×80, H&E

corticosterone maintains a high ACTH level which drives the impaired metabolism in the cortical cells.

Pigmentation (brown atrophy)

Pigment (ceroid or lipofuscin) may be seen in the cells of the zona reticularis in the aged AP rat, and it has been recorded in several other strains (Greaves and Faccini, 1984; Parker and Valerio, 1983). The pigment is thought to be the result of peroxidation of unsaturated fat and may be enhanced in deficiencies of vitamin E and other antioxidants.

Atrophy

Atrophy of the inner two zones of the adrenal cortex has been observed in AP rats with cortical neoplasms in the contralateral gland, and in a few animals with large pituitary tumours. The histological appearance is of a thin cortex where the glomerulosa cells appear relatively normal but the remaining cells are small and depleted of lipid. The capsule is usually greatly thickened. The assumption is that ACTH secretion has been suppressed, in one case by the secretion of cortico-sterone from the cortical tumour, and in the other by loss of pituitary ACTH-secreting cells due to compression or invasion from the pituitary tumour.

Cellular alteration/hyperplasia

This term is applied to small groups of cells which differ from surrounding cells in cytology. In the cortex, altered cells may be larger than surrounding cells, and their nuclei may also be enlarged. The cytoplasm may contain more lipid, or be more eosinophilic or basophilic. As the foci enlarge they may merit the description hyperplastic nodule as they become more nodular and show some compression (Figure 100). In BN/Bi rats the incidence of cortical cellular alteration may reach levels of 100 per cent (Burek, 1978). In the AP rat, levels up to 5 per cent for altered foci and 10 per cent for hyperplastic nodules have been recorded in 2 year studies.

Proliferative lesions of the adrenal medulla are a frequent finding in the rat, in contrast to humans and other laboratory species. The hyperplasia involves both adrenaline- and noradrenaline-secreting cells (Tischler *et al.*, 1985). Diffuse medullary hyperplasia occurs when the medullary cells are increased without nodule formation or cortical compression. It is a bilateral and multicentric condition, and in male Wistar rats aged 2 years the adrenaline-producing cells are increased by 40 per cent and noradrenaline cells by 60 per cent. Nodular hyperplasia of the medulla is considered to be part of a continuous process proceeding to phaeochromocytoma, and the distinction between hyperplasia and neoplasia remains a subject of debate. Nodular hyperplasia may occur in any part of the medulla, but the juxta-cortical position is the most frequent site in the AP rat. The cells are smaller than normal, the

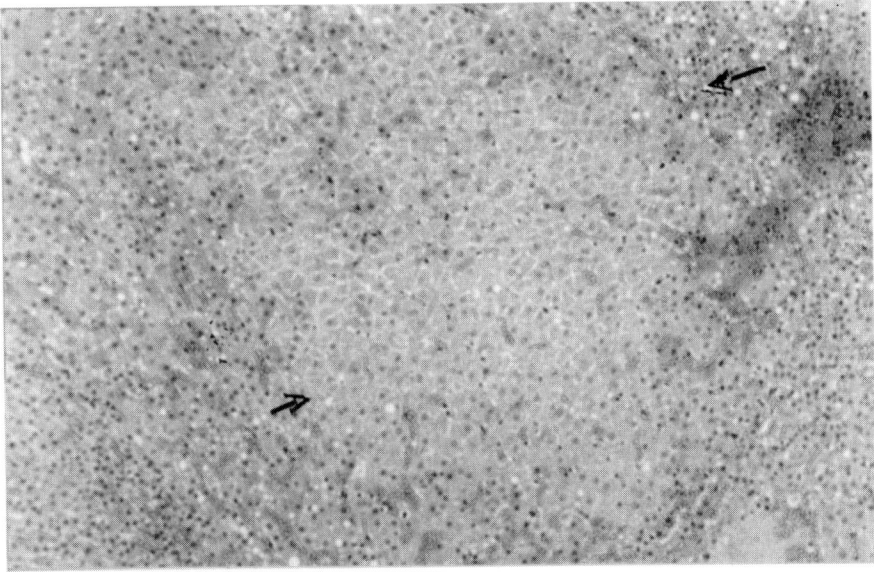

Figure 100   Adrenal cortex of a male AP rat aged 26 months. Area of cellular alteration (→). ×80, H&E

cytoplasm more basophilic, and mitotic figures may be present. The distinction from neoplasia, which has been used in AP rats, is that the focus does not show compression of medulla or cortex, even when the nodule is large and there is not a clear margin. Medullary hyperplasia is common in old AP rats, particularly males, and there appears to be some association with renal disease. It has been shown by Roe and Bär (1985) that medullary hyperplasia can be induced by interfering with calcium homeostasis. Chronic progressive glomerulonephropathy in the rat impairs the animal's ability to handle calcium overload, so that this may be the mechanism for the induction of medullary hyperplasia.

11.3.2 Neoplastic Changes

Cortex

Cortical tumours include adenomas and carcinomas. Cortical adenomas may resemble a hyperplastic nodule except that there is a clear margin to the tumour and adjacent cells are compressed (Figures 101 and 102); capsules are not common. Carcinomas show more striking cytological differences. The cells are larger and pleomorphic and the cytoplasm is eosinophilic, often with less lipid. The architecture of the carcinomas is variable, with the formation of cords, sheets or nodules. Necrosis and haemorrhage are common features and mitotic figures may be numerous. Local invasion of capsule and other tissues is present, but metastases are not common.

199

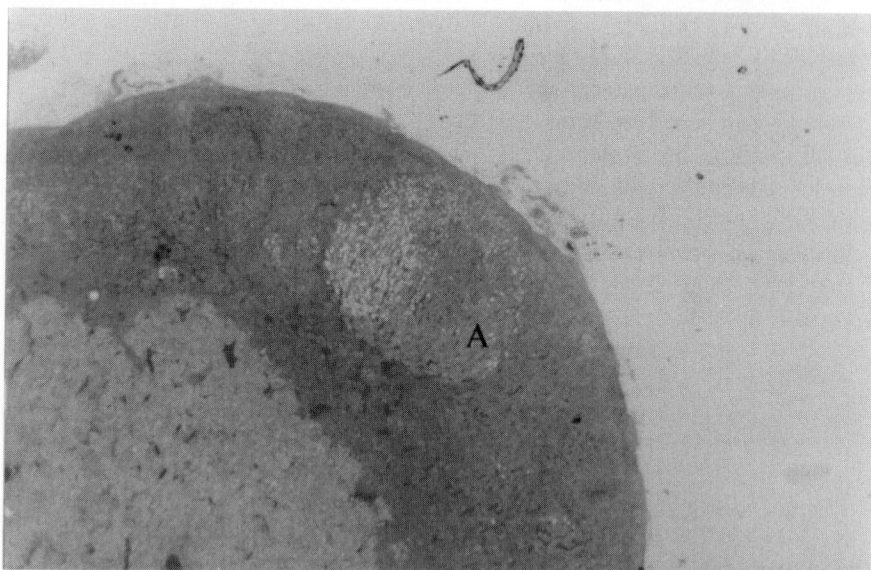

Figure 101   Small cortical adenoma (A) in a male AP rat aged 20 months. ×8, H&E

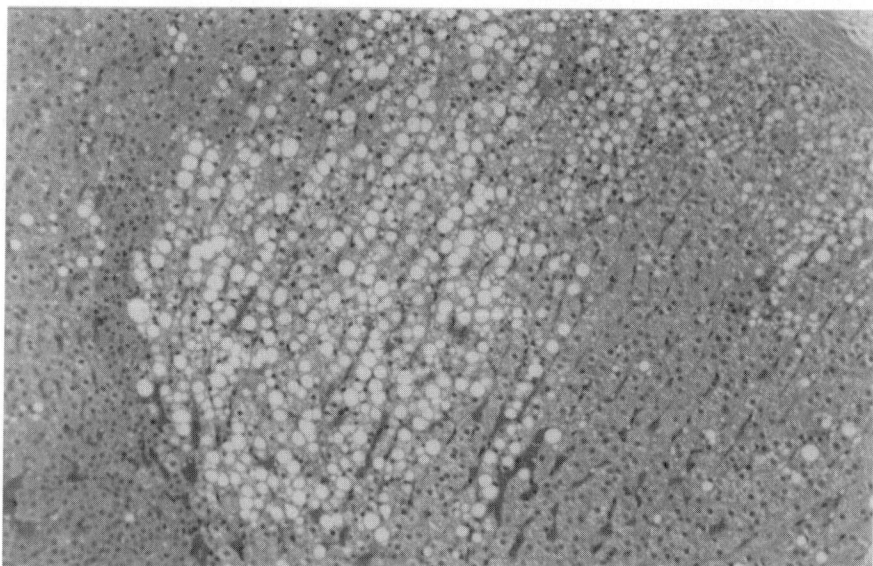

Figure 102   Higher power view of cortical adenoma showing compression of cells at the margin. ×32, H&E

The incidence of cortical tumours in the AP rat is low, and they were mostly adenomas with a few carcinomas, two of which had lung metastases. In the majority of 2 year studies there were no tumours of the adrenal cortex. In the life-span study the incidence was 3 per cent in males, 1 per cent in females, and all of the tumours were found in animals more than 2 years old. In standard carcinogenicity tests when cortical tumours were present there was a slight preponderance in males, with the highest incidence in control animals of 2 per cent. Low incidences (<10 per cent) have also been reported for the SD and F344 strains (Suzuki *et al.*, 1979; Goodman *et al.*, 1979; Maekawa *et al.*, 1983). By contrast the Osborne-Mendel rat has the highest levels: up to 65 per cent at 18 months (Hollander and Snell, 1976).

Medulla

The most common tumour of the adrenal medulla of the rat is the phaeochromocytoma, which may be benign or malignant and is usually unilateral. The histological appearance of the tumour is of small cells with some nuclear pleomorphism, increased cytoplasmic basophilia, compression of adjacent tissue, and a clear margin. Atypia and mitotic figures may be a feature of the malignant tumours. In AP rats the incidence is higher in males and, in 2 year studies, ranges between 2 and 6 per cent in males and 0 to 3 per cent in females. There has been no significant change in the incidence of phaeochromocytomas in the AP rat in the last thirty years. Low incidences have been reported for the Osborne-Mendel rat (Goodman *et al.*, 1980) but Tannenbaum *et al.* (1962) reported moderate levels in SD rats (males 31 per cent and females 5 per cent), and similar incidences were reported by Solleveld *et al.* (1984) in F344 rats (males 30 per cent, females 15 per cent).

A relationship between phaeochromocytoma and renal disease in the rat was first observed by Gilman *et al.* (1953). The mechanism for induction of tumours, as for medullary hyperplasia, is likely to be effects on calcium homeostasis. Phaeochromocytomas in the rat do show increased production of noradrenaline (Bosland and Baer, 1984).

Ganglioneuromas of the adrenal medulla are rare tumours characterised by the presence of ganglion cells, although chromaffin cells may be present in the neuromatous stroma. Ganglioneuromas in the AP rat have been reviewed by Glaister *et al.* (1977). The only other primary tumour of the adrenal seen in the AP rat was a fibrosarcoma arising in the capsule. Secondary infiltration of the adrenals by lymphomas and leukaemias has been observed and metastases of an osteosarcoma.

11.4 Thyroid Gland

The thyroid gland is located below the cricoid cartilage and is involved with the maintenance of metabolic functions throughout the body through the hormones triiodothyronine (T_3) and tetraiodothyronine (thyroxine, T_4), and unlike other endocrine glands the thyroid has a large storage capacity. Its hormone production is entirely dependent on the availability of iodine, which can vary widely in different areas. The iodide content of drinking water varies in different geographical areas, and the levels in laboratory diets are dependent on the source of protein. In standard toxicology studies in the AP rat the thyroid gland is removed from the animal, attached to the trachea, and is not weighed. In our laboratory it is considered that to provide accurate weights by complete removal of the two lobes and isthmus is too difficult to be accurate, and causes unacceptable damage to the histology of the thyroid. If there are indications of effects in the thyroid gland, separate investigative studies are done, and the thyroid glands are normally weighed in this type of study. Limited data on the levels of thyroid hormones in the AP rat are given in Chapter 1 (Table 1.11) and show that while TSH and T_3 levels are similar in males and females, levels of thyroxine in males are approximately twice those in females. The levels are, in general, similar to those reported for SD, Osborne-Mendel and Long-Evans rats (Gregerman and Crowder, 1963; Tang, 1985; Donda *et al.*, 1987). The thyroid weight in the rat increases from 3 to 6 mg in the first week of life and stabilises at about 20 mg at 20 weeks of age (Thomas and Williams, 1994).

11.4.1 Structure and Function

Follicular cells

The follicles of the thyroid gland are closed sacs containing a colloidal material of stored hormone, surrounded by a single layer of the main epithelial component of the gland, the follicular cell. These cells have microvilli and pseudopods at the apical surface which have great importance in the secretory activity of the follicle. There is a stroma of fibroblasts and endothelial cells to which the basal portion of the follicular cell is closely apposed. The follicles are arranged in lobules, each with a terminal branch of the vascular system. There is regional variation in the size of the follicles with smaller follicles at the centre of the gland and larger follicles lined by flatter cells at the periphery (Figure 103). There are also sex differences in the height of the follicular epithelium; in general males always have a greater epithelial height due to higher levels of thyroid stimulating hormone (TSH) and a higher metabolic rate (Doniach, 1969).

Iodine is taken into the follicular cells by a pump mechanism which is controlled by TSH, oxidised and transported to the apical border. Thyroglobulin is synthesised in the ribosomes of the endoplasmic reticulum of the follicular cells and passed to the lumen of the follicle where iodination occurs and T_3 and T_4 are formed. The hormones are stored in the colloid until required. On receipt of a stimulus by TSH the microvilli at the apical side of the follicular cell extend in to the colloid and ingest a fragment, a process termed

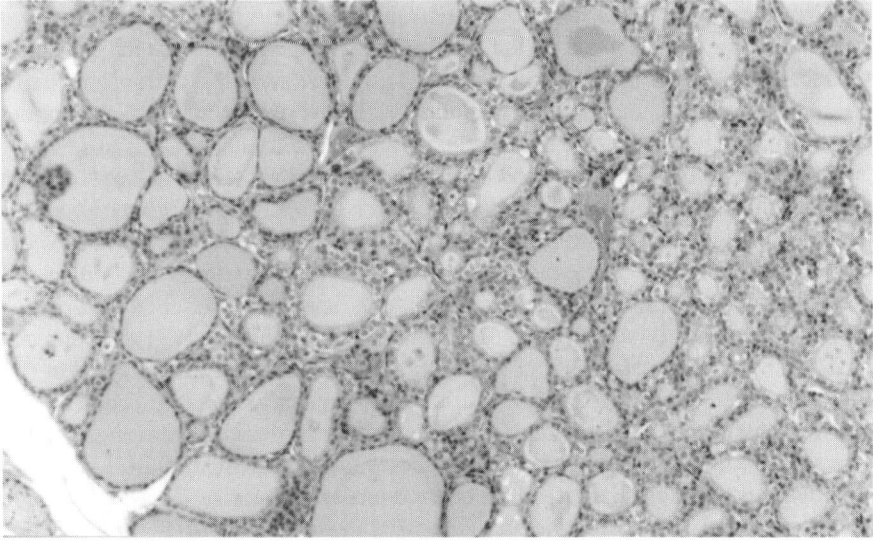

Figure 103 Thyroid of a male AP rat aged 3 months showing small follicles located towards the centre and larger follicles at the periphery. ×80, H&E

endocytosis. Proteolytic enzymes release T_3 and T_4 which diffuse out of the cells into the rich network of interfollicular capillaries. More T_4 is produced than T_3 and the latter can be formed by deiodination of T_4. Circulating T_3 and T_4 are mostly protein bound and degraded, by conjugation, in the liver and excreted into the bile. The plasma half-life of T_4 is short in the rat (up to 24 hours) compared with man (up to 9 days). The difference is due to the presence in humans of the protein thyroxine-binding-globulin which has a high affinity for thyroxine; it is not present in the rat. Control of thyroid function is modulated by TSH which is itself controlled by thyrotropin-releasing hormone (TRH) from the hypothalamus. The thyrotrophic cells of the anterior pituitary have nuclear receptors for T_3 and, when the level of binding drops, release of TSH occurs. TSH increases the uptake of iodine into the follicular cell and also increases transcription of thyroglobulin and thyroid peroxidase genes. Since T_4 is deiodinated to T_3 a drop in the circulating level of thyroxine, caused by decreased synthesis or increased clearance, will also reduce the amount of T_3. The level of thyroxine is normally relatively constant, although TSH has a diurnal secretion. This is the very sensitive feedback mechanism which maintains normal thyroid follicular function.

C-cells (parafollicular cells)

In the rat C-cells are located at the basal region of the follicular cells. They have a polygonal or spindle shape with clear, finely granular cytoplasm and large centrally located nuclei. They have an intense argyrophilia but are best

203

demonstrated by immunocytochemical staining for calcitonin (CT), which is the chief secretory product of the cell. Ultrastructural examination shows the presence of abundant membrane-bound secretory granules in most C cells, although a few may be sparsely granulated. C-cells also contain other peptides including somatostatin and gastrin-releasing peptide.

CT is involved in the control of calcium and phosphorus in concert with parathyroid hormone and interacts with target cells in bone, kidney and, to a lesser extent, intestine. It inhibits bone resorption and consequently the release of calcium from bone, but decreases renal tubular absorption of phosphate. CT and parathyroid hormone together provide a dual negative feedback which maintains calcium levels within narrow margins. Kalu *et al.* (1983) showed that in the F344 rat serum CT concentration increases from less than 90 pg/ml in 6 week old rats to 14 ng/ml in rats over two years of age while serum calcium levels show little change over the same time period. Kalu *et al.* (1983) also showed that food restriction diminishes this increase in serum CT levels. It is suggested that *ad libitum* feeding could cause an excessive intake of calcium from laboratory animal diets which are calcium rich. Any factor, such as this, which predisposes to hypercalcaemia could stimulate CT secretion.

11.4.2   *Non-neoplastic Changes*

Congenital anomalies

Ectopic thymus and ultimobranchial cysts, lined by a squamous epithelium, are not uncommon findings in the AP rat.

Inflammation

Inflammation of the thyroid gland is rare in the AP rat and most other strains. Small focal inflammatory cell infiltrates, usually of mononuclear cells, are seen frequently in the thyroid, but they are considered to be of no biological significance. The Buffalo and BioBreeding/Worcester rat strains both develop a spontaneous lymphocytic thyroiditis with age (Silverman and Rose, 1974; Yanagisawa *et al.*, 1986). Whole body irradiation and thymectomy can induce chronic thyroiditis in strains which do not develop the condition spontaneously (Penhale *et al.*, 1973), which suggests that they do not develop the immuno-suppression which causes the disease.

Pigmentation

Aging AP rats show an increase in iron and lipofuscin in the follicular epithelium and also in the colloid. This has been reported in other strains (Ward and Reznik-Schüller, 1980). Among SD rats pigmentation has been observed more frequently in SD CD rats rather than SD/HAP rats (Anver *et al.*, 1982). In F344 rats the incidence reported by Goodman *et al.* (1979) was low, with pigmentation in <1 per cent of animals.

Cystic follicles

Cystic follicles are seen occasionally in aging AP rats. They are located at the periphery of the gland and are more common in females. Histologically they are very large follicles, filled with pale colloid and lined by a flattened epithelium.

Hypertrophy/hyperplasia

Increased functional demand in the thyroid gland can result in cellular hypertrophy as well as hyperplasia. The follicular epithelium will increase in height in response to TSH and there is frequently an increase in the basophilic staining of the colloid in H&E sections. This diffuse hyperplasia appears as a uniform increase in the size of the follicles. The epithelium of the follicles is columnar with large basally located nuclei (Figure 104) and may be stratified or papillary. Nodular follicular cell hyperplasia in the thyroid gland has similar characteristics to hyperplasia in other endocrine glands in that there is not usually compression of adjacent tissue, capsule formation or marked cellular atypia. The hyperplastic areas show follicles with an infolded papillary epithelium, and in long-standing lesions haemorrhage, fibrosis, pigmentation and cholesterol clefts may be present. C-cell hyperplasia can also be diffuse

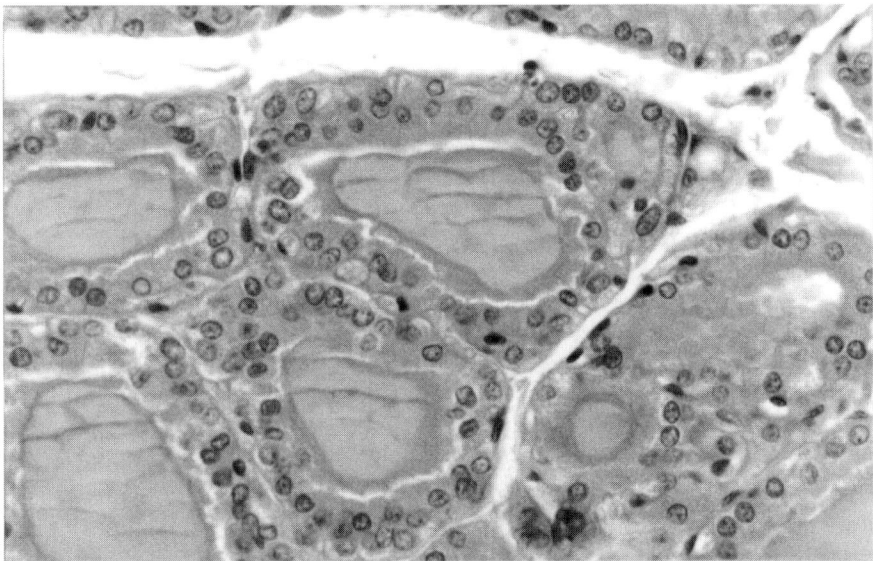

Figure 104 Diffuse hyperplasia of the thyroid: follicles have a columnar epithelium with basally located nuclei. ×128, H&E

with increased numbers of cells throughout the gland (Williams, 1966; DeLellis *et al.*, 1979). In the AP rat diffuse hyperplasia is not as common as nodular hyperplasia. In this type of hyperplasia small lobules of C-cells, the size of one to two follicles, are formed. The C-cells within the nodules are uniform and essentially normal in appearance. As the foci enlarge, atrophic follicles may be found within the lobule. Immunostaining of nodular hyperplastic C-cells shows a variable immunoreactivity for CT, whereas the cells in diffuse hyperplasia tend to demonstrate generalised high levels of the hormone.

11.4.3 Neoplastic Changes

Follicular cell tumours

Adenomas of follicular cells in the AP rat have been diagnosed on the basis of the following criteria: the tumours are single and encapsulated with clear compression of surrounding tissue. The architecture may be of follicles of variable size, but of relatively normal appearance, although the cell cytoplasm often stains more deeply basophilic than surrounding follicles. Less frequently the follicles have a florid papillary epithelium, and the least common subtype is a solid mass of undifferentiated cells. This latter type should only be diagnosed in the absence of any immunoreactivity for CT. Malignant follicular tumours have a similar histological appearance to the benign tumour but show distinct invasion through the capsule into adjacent tissues. Lung metastases have been observed in only one animal. With the exception of the few malignant tumours, the adenomas were incidental findings at necropsy or microscopic examination.

C-cell adenomas share many of the features of nodular hyperplasia, which probably represents a stage in the progression to tumour. The difference from hyperplasia is the larger size of the adenoma and the presence of compression. Capsules are not usually evident and, as for follicular tumours, the majority were incidental findings during microscopic examination. Malignant C cell tumours show local invasion, a greater degree of cellular and nuclear atypia, and mitoses may be abundant. Metastases to deep cervical lymph nodes have been seen in occasional animals and this was also reported by Burek (1978) in 10 per cent of a Wistar-derived strain.

The incidence of follicular and C-cell tumours in long term studies in the AP rat is shown in Table 11.5. Since the diagnostic criteria for thyroid tumours have changed in the last thirty years the tumours from all studies prior to 1985 were re-examined and, if necessary, reclassified. In the life-span study only two tumours were seen in animals dying before 28 months of age. There appears to be an increase in the incidence of spontaneous follicular tumours with time and they are more common in males. C-cell tumours are slightly more frequent in females but there has been no significant change in the incidence level with time. A single squamous carcinoma, probably

Table 11.5 Incidence of thyroid tumours in the AP rat

	Follicular tumours			C-cell tumours		
Year[a]	Males		Females	Males	Females	
1963 (Life-span)[b]	1/194 (<1%)		1/302 (<1%)	4/194 (2%)	10/302 (3%)	
1969	0/20		0/20	2/20 (10%)	0/20	
1973	0/25		0/25	2/25 (8%)	0/25	
1974	0/50		0/50	2/50 (4%)	2/50 (4%)	
1975	1/65 (2%)		0/65	3/65 (5%)	4/65 (6%)	
1976	0/65		0/65	4/65 (6%)	5/65 (7%)	
1977	1/30 (3%)		2/30 (6%)	2/30 (6%)	4/30 (13%)	
1982	7/150 (5%)		5/150 (3%)	6/150 (4%)	9/150 (6%)	
1986	7/150 (5%)		0/150	5/150 (3%)	8/150 (5%)	
1989	5/50 (10%)		1/50 (2%)	3/50 (6%)	6/50 (12%)	

[a] The year the study was completed
[b] The life-span study ran for 52 months; all other studies were of 24 months duration and only those studies are included where it was possible to review all the thyroid tumour slides.

originating in an embryonic cyst, has been observed. Follicular adenomas are uncommon in most strains of rat. In the F344, levels less than 1 per cent have been recorded (Jacobs and Huseby, 1967; Goodman *et al.*, 1979; Solleveld *et al.*, 1984), and in the SD rat even lower levels have been recorded (Mackenzie and Garner, 1973). The Osborne-Mendel rat and the Long-Evans have a slightly higher incidence, around 3 per cent (Goodman *et al.*, 1980; Lee *et al.*, 1982).

There are also strain differences in the incidence of C-cell tumours in rats, although reported figures must be regarded with caution since there are significant differences in the criteria used for distinguishing hyperplasia from tumours. Also in older studies before CT immunostaining was available the distinction between solid follicular cell tumours and C-cell tumours was difficult. C-cell tumours are common in Long-Evans and WAG/Rij strains and F344 rats (Burek, 1978; DeLellis *et al.*, 1979; Goodman *et al.*, 1979; Maekawa *et al.*, 1983; Solleveld *et al.*, 1984) and lower in Osborne-Mendel and SD rats (Mackenzie and Garner, 1973).

Dietary factors may be involved in the development of thyroid follicular tumours in the rat as prolonged exposure to iodide deficiency or dietary goitrogens can cause thyroid tumours. Among dietary goitrogens are the thiocyanates in cassava, thiourea derivatives in brassicas, and goitrogens in soya. Goitrogenic substances have also been identified in the water. Increased food intake has been associated with increased C-cell tumours as has increasing levels of vitamin D_3 (Thurston and Williams, 1982).

Thyroid follicular tumours have been induced in the AP rat by several drugs with widely different pharmacological actions. In all cases tumours were slow to develop, were more common in males and were preceded and accompanied by thyroid hypertrophy and hyperplasia.

11.5 Parathyroid Gland

The rat has a single pair of parathyroid glands located with the thyroid gland; since they are not in similar positions within the thyroid lobes they are not uniformly sectioned in standard sections. In toxicology studies in the AP rat it is a requirement that one of the two glands must always be present in the section of thyroid. Histologically the glands are composed of a single cell type, the chief cell, which produces only the parathyroid hormone (PTH). The majority of cells at any one time are inactive. When serum calcium levels are low an active phase of hormone synthesis ensues after which the chief cells return to the resting phase. Serum levels of calcium, and to a lesser extent magnesium, are the feedback mechanisms for controlling PTH synthesis. Chief cells also secrete chromogranin A, a protein which is stored and secreted with PTH. The function of PTH is to affect target cells in the bone, kidney and intestine to maintain calcium homeostasis. PTH mobilises bone calcium and increases renal tubular reabsorption of calcium.

11.5.1 *Non-neoplastic Changes*

Cysts/inflammation/necrosis

Occasional blood- or serous-fluid-filled cysts have been seen in the parathyroid glands of the AP rat. Both necrosis and inflammation are rare conditions but have been seen in a few animals.

Hyperplasia

Two types of hyperplasia have been seen in the AP rat. Diffuse hyperplasia secondary to severe renal disease (Figure 105) is seen, chiefly in males over 18 months of age. The enlargement is usually visible macroscopically and is due to both hypertrophy and hyperplasia of the chief cells. The cells may have a slightly eosinophilic cytoplasm and vacuoles may be present. An increased fibrous stroma may give the glands a lobulated appearance. These enlarged parathyroid glands produce excessive amounts of PTH. Focal hyperplasia of chief cells is very much less common in the AP rat (<0.1 per cent). The focal hyperplastic area does not compress adjacent cells and is not encapsulated. The chief cells have a more extensive cytoplasm which confers a paler appearance than normal cells.

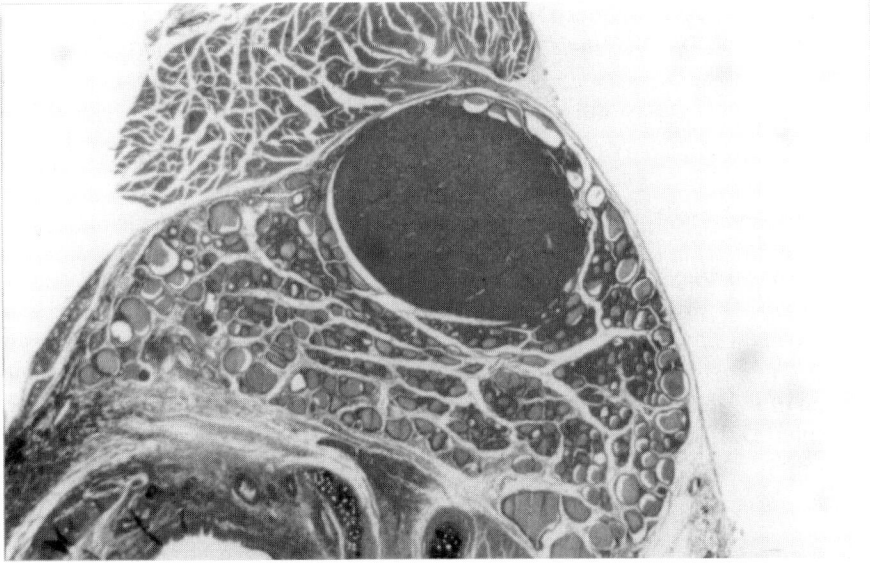

Figure 105 Diffuse hyperplasia of the parathyroid in a 26 month old AP rat with severe renal disease. ×8, H&E

11.5.2 Neoplastic Changes

Only two adenomas of the parathyroid gland have been observed in the AP rat. They were both large tumours, occupying approximately half of the gland. They had a clear margin and were partly encapsulated, but the cells showed little difference from the normal chief cell and only a few mitotic figures were observed. Parathyroid tumours are rare in all rat strains and are considered to be non-functional (Capen, 1994). The marked, diffuse hyperplasia, secondary to renal disease, does not appear to be a pre-neoplastic change in the gland.

11.6 References

ANVER, M. R., COHEN, B. J., LATTUDA, C. P. and FOSTER, S. J. (1982) Age-associated lesions in barrier reared male Sprague-Dawley rats: a comparison between HAP:(SD) and Crl:Cobs, CD (SD) stocks, *Experimental Aging Research*, **8**, 3–24.

BARSOUM, N. J., MOORE, J. D., GOUGH, A. W., STURGESS, J. M. and DE LA IGLESIA, F. A. (1985) Morphofunctional investigations on spontaneous pituitary tumours in Wistar rats, *Toxicologic Pathology*, **13**(3), 200–8.

BERKVENS, J. M., VAN NESSELROOY, J. H. J. and KROES, R. (1980) Spontaneous tumours in the pituitary gland of old Wistar rats. A morphological and immunocytochemical study, *Journal of Pathology*, **130**, 179–91.

BERRY, P. H. (1986) Effects of diet or reproductive status on the histology of spontaneous pituitary tumours in female Wistar rats, *Veterinary Pathology*, **23**, 606–18.

BOSLAND, M. C. and BAER, A. (1984) Some functional characteristics of adrenal medullary tumours in aged male Wistar rats, *Veterinary Pathology*, **21**, 129–40.

BUREK, J. D. (1978) *Pathology of Aging Rats*, pp. 54–7, West Palm Beach, Florida: CRC Press.

CAPEN, C. C. (1994) Changes in the structure and function of the parathyroid gland, in MOHR, U., DUNGWORTH, D. L. and CAPEN, C. C. (Eds), *Pathobiology of the Aging Rat*, pp. 197–226, Washington: ILSI Press.

CARLTON, W. W. and GRIES, C. L. (1983) Cysts, pituitary, rat, mouse, hamster, in JONES, T. C., MOHR, U. and HUNT, R. D. (Eds), *Endocrine System*, pp. 161–3, Berlin: Springer-Verlag.

CHAUDHURI, M., SARTIN, J. L. and ADELMAN, R. C. (1983) A role for somatostatin in the impaired insulin secretory response to glucose by islets from aging rats, *Journal of Gerontology*, **38**, 431–5.

CHILDS, G. V., ELLISON, D. G. and GARNER, L. L. (1980) An immunocytochemist's view of gonadotrophin storage in the adult male rat: cytochemical and morphological heterogeneity in serially sectioned gonadotropes, *American Journal of Anatomy*, **158**, 397–409.

COLEMAN, G. L., BARTHOLD, S. W., OSBALDISTON, G. W., FOSTER, S. J. and JONAS, A. M. (1977) Pathological changes during aging in barrier-reared Fischer 344 male rats, *Journal of Gerontology*, **32**, 258–78.

CONYBEARE, G. (1988) Modulating factors: challenges to experimental design, in GRICE, H. C. and CIMINERA, J. L. (Eds), *Carcinogenicity*, Washington: ILSI Press.

DeLELLIS, R. A., NUNNEMACHER, G., BITMAN, W. R. and GAGEL, R. F. (1979) C-cell hyperplasia and medullary thyroid carcinoma in the rat. An immunohistochemical and ultrastructural study, *Laboratory Investigation*, **40**, 140–54.

DHOM, G., HOHBACH, C., MAUSLE, E. and SCHERR, O. (1981) Peliosis of the female adrenal cortex of the aging rat, *Virchows Archiv [A]*, **36**, 195–206.

DODD, D. C., PORT, C. D., DESLEX, P., REGNIER, B., SANDERS, P and INDACOCHEA-REDMOND, N. (1987) Two-year evaluation of misoprostol for carcinogenicity in CD Sprague-Dawley rats, *Toxicologic Pathology*, **15**(2), 125–33.

DONDA, A., REYMOND, M. J., ZURICH, M. G. and LEMARCHAND-BERAUD, T. H. (1987) Influence of sex and age on T3 receptors and T3 concentration in the pituitary gland of the rat: consequences on TSH secretion, *Molecular and Cellular Endocrinology*, **54**, 29–34.

DONIACH, I. (1969) Correlation of thyroid cell height with sex differences in tumor induction: discussion on carcinogenic role of TSH, in HEIDINGER, C. E. (Ed.), *Thyroid Cancer*, pp. 131–4, New York: Springer-Verlag.

DRAZNIN, B., STEINBERG, J. P., LEITNER, J. W. and SUSSMAN, K. E. (1985) Nature of insulin secretory defect in aging rats, *Diabetes*, **34**, 1168–73.

DRIBBEN, J. S. and WOLFE, J. M. (1947) Structural changes in the connective tissue of the adrenal glands of female rats, *Anatomical Record*, **98**, 557–85.

FURTH, J., CLIFTON, K. H., GODSDEN, E. L. and BUFFETT, R. F. (1956) Dependant and autonomous mammotropic pituitary tumors in rats: their somatotropic features, *Cancer Research*, **16**, 608–15.

FURTH, J., NAKANE, P. and PASTEELS, J. L. (1976) Tumours of the pituitary gland, in TURUSOV, V. S. (Ed.), *Pathology of Tumours in Laboratory Animals,* Vol. 1, *Tumours of the Rat,* pp. 201–37, Lyon: IARC.

GILMAN, J., GILBERT, C. and SPENCE, I. (1953) Phaeochromocytoma in the rat. Pathogenesis and collateral reactions and its relation to comparable tumours in man, *Cancer,* **6,** 494–511.

GLAISTER, J. R., SAMUELS, D. M. and TUCKER, M. J. (1977) Ganglioneuroma-containing tumours of the adrenal medulla in Alderley Park rats, *Laboratory Animals,* **11,** 35–7.

GOODMAN, D. G., WARD, J. M., SQUIRE, R. A., CHU, K. C. and LINHART, M. S. (1979) Neoplastic and non-neoplastic lesions in aging F344 rats, *Toxicology and Applied Pharmacology,* **48,** 237–48.

GOODMAN, D. G., WARD, J. M., SQUIRE, R. A., CHU, K. C., and LINHART, M. S. (1980) Neoplastic and non-neoplastic lesions in aging Osborne-Mendel rats, *Toxicology and Applied Pharmacology,* **55,** 433–47.

GOSNEY, J. A. (1986) Morphological changes in the pituitary and thyroid of the rat in hypobaric hypoxia, *Journal of Endocrinology,* **109,** 119–24.

GREAVES, P. and FACCINI, J. M. (1984) Endocrine glands, in *Rat Histopathology: A glossary for use in toxicity and carcinogenicity studies,* pp. 187–210, Amsterdam: Elsevier.

GREENHILL, R. (1992) The pituitary gland in different species, in ATTERWILL, C. K. and FLACK, J. D. (Eds), *Endocrine Toxicology,* pp. 15–49, Cambridge: Cambridge University Press.

GREGERMAN, R. I. and CROWDER, S. E. (1963) Estimation of thyroxine secretion rate in the rat by the radioactive thyroxine turnover technique: influences of age, sex and exposure to cold, *Endocrinology,* **72,** 383–92.

HADJU, A. and RONA, G. (1967) Morphological observations on spontaneous pancreatic islet changes in rats, *Diabetes,* **16,** 108–10.

HINSON, J. P. and RAVEN, P. W. (1996) Adrenal morphology and hormone synthesis and regulation, in HARVEY, P. W. (Ed.), *The Adrenal in Toxicology, Target Organ and Modulator of Toxicity,* pp. 23–52, London: Taylor & Francis.

HOLLANDER, C. F. and SNELL, K. C. (1976) Tumours of the adrenal gland, in TURUSOV, V. (Ed.), *Pathology of Tumours in Laboratory Animals,* Vol. 1, *Tumours of the Rat,* pp. 273–94, Lyon: IARC.

IBRAHIM, S. N., MOUSSA, S. M. and CHILDS, G. V. (1986) Morphometric studies of rat anterior pituitary cells after gonadectomy: correlation of changes in gonadotrophs with the serum levels of gonadotrophins, *Endocrinology,* **119,** 629–37.

ITO, I., SATO, Y. and SAZUKI, H. (1986) Increase in adrenal catecholamine secretion and adrenal sympathetic nerve unitary activities with aging in rats, *Neuroscience Letters,* **69,** 263–8.

JACOBS, B. B. and HUSEBY, R. A. (1967) Neoplasms occurring in aged Fischer rats with special reference to testicular, uterine and thyroid tumours, *Journal of the National Cancer Institute,* **39,** 303–9.

KALU, D., COCKERHAM, R., YU, B. P. and ROOS, B. A. (1983) Lifelong dietary modulation of calcitonin levels in rats, *Endocrinology,* **113,** 2010–16.

KROES, R., GARBIS-BERKVENS, J. M., DE VRIES, T. and VAN NESSELROOY, J. H. (1981) Histopathological profile of a Wistar rat stock including a survey of the literature, *Journal of Gerontology,* **36,** 259–79.

Diseases of the Wistar Rat

Diseases of the Wistar Rat

KVETNANSKY, R., WEISE, V. K., GEWIRTZ, G. P. and KOPIN, I. J. (1971) Synthesis of catecholamines in rats during and after immobilization stress, *Endocrinology*, **89**, 46–9.

LECHAN, R. M. (1987) Neuroendocrinology of pituitary hormone regulation, *Endocrinology and Metabolism Clinics*, **16**(3), 475–501.

LEE, W., CHIACCHIENNI, R. P., SHELEIN, B and TELLES, N. C. (1982) Thyroid tumors following ^{131}I or localised X-irradiation to the thyroid and pituitary glands in rats, *Radiation Research*, **92**, 307–19.

LLOYD, R. V. (1983) Estrogen induced hyperplasia and neoplasia in the rat anterior pituitary gland, *American Journal of Pathology*, **113**, 198–206.

LLOYD, R. V. and MAILLOUX, J. (1987) Effects of diethylstilboestrol and propyl-thiouracil on the rat pituitary, *Journal of the National Cancer Institute*, **79**, 865–73.

LUMB, G., MITCHELL, L. and DE LA IGLESIA, F. A (1985) Regression of pathological changes induced by long-term administration of contraceptive steroids to rodents, *Toxicologic Pathology*, **13**, 283–95.

MACKENZIE, W. F. and GARNER, F. M. (1973) Comparison of tumours in six sources of rats, *Journal of the National Cancer Institute*, **50**, 1243–57.

MAEKAWA, A., KUROKAWA, Y., TAKAHASHI, M., KUKUBO, T., OGUI, T., ONODERA, H., TANIGAWA, H., OHNO, Y., FURUKAWA, F. and HAYASHI, Y. (1983) Spontaneous tumors in F344/DuCrj rats, *Gann*, **74**, 365–72.

MAGNUSSON, G., MAJEED, S. K. and GOPINATH, C. (1979) Infiltrating pituitary neoplasia in the rat, *Laboratory Animals*, **13**, 111–13.

MCCARTY, R. and KOPIN, I. J. (1978) Sympatho-adrenal medullary activity and behaviour during exposure to footshock stress: a comparison of seven rat strains, *Physiology and Behaviour*, **21**, 567–72.

MCCOMB, D. J., KOVACS, K., BERI, J. and ZAK, F. (1984) Pituitary adenomas in old Sprague-Dawley rats; a histologic, ultrastructural and immunocytochemical study, *Journal of the National Cancer Institute*, **73**(5), 1143–66.

MOLINA, J. M., PREMDAS, F. H., LIPSON, L. G. (1985) Insulin release in aging: dynamic response of isolated islets of Langerhans of the rat to D-glucose and D-glyceraldehyde, *Endocrinology*, **116**, 821–6.

MUNCK, A., GUYRE, P. M. and HOLBROOK, N. J. (1984) Physiological functions of glucocorticoids in stress and their relation to pharmacological actions, *Endocrine Reviews*, **5**, 25–44.

NUSSDORFER, G. G. (1980) Cytophysiology of the adrenal cortex, *International Reviews of Cytology*, **64**, 307–69.

OLDSTONE, M. B. A., SINHA, Y. N., BLOUNT, P., TISHON, A., RODRIGUEZ, M., VON WEDEL, R. and LAMPERT, P. W. (1982) Virus alterations in homeostasis: alterations in differentiated functions of infected cells in vivo, *Science*, **218**, 1125–7.

PARKER, G. A. and VALERIO, M. G. (1983) Lipogenic pigmentation, adrenal cortex, rat, in JONES, T. C., MOHR, U. and HUNT, R. D. (Eds.), *Endocrine System*, pp. 64–6, Berlin: Springer-Verlag.

PENHALE, W. J., FARMER, A., MCKENNA, R. P. and IRVINE, W. J. (1973) Spontaneous thyroiditis in thymectomised and irradiated Wistar rats, *Clinical and Experimental Immunology*, **15**, 225–36.

PICKERING, C. E. and PICKERING, R. G. (1984) The effect of repeated reproduction on the incidence of pituitary tumours in Wistar rats, *Laboratory Animals*, **18**, 371–8.

212

PRYSOR-JONES, R. A., SILVERLIGHT, J. J. and JENKINS, J. S. (1983) Hypothalamic dopamine and catechol oestrogens in rats with spontaneous pituitary tumours, *Journal of Endocrinology*, **96**, 347–52.

PURVES, H. D. and GRIESBACH, W. E. (1956) Changes in the basophil cells of the rat pituitary after thyroidectomy, *Journal of Endocrinology*, **13**, 365–75.

REAVEN, E., KOSTRNA, M., RAMACHANDRAN, J. and AZHAR, S. (1988) Structure and changes in the adrenal glands during aging, *American Journal of Physiology*, **255**, 903–11.

REBUFFAT, P., BELLONI, A. S., ROCCO, S., ANDREIS, P. G., MALENDOWICZ, L. K., GOTTARDO, G., MAZZOCCHI, G. and NUSSDORFERM, G. G. (1992) The effects of aging on the morphology and function of the zonae fasciculata/reticularis of the rat adrenal cortex, *Cell and Tissue Research*, **270**, 265–72.

REICHLIN, S. (1989) Neuroendocrinology of the pituitary gland, *Toxicologic Pathology*, **17**(2), 250–5.

RILEY, M. G. I. and BOORMAN, G. A. (1990) Endocrine pancreas, in BOORMAN, G. A., EUSTIS, S. L., ELWELL, M. R. and MONTGOMERY, C. A., *Pathology of the Fischer Rat*, pp. 545–53, New York: Academic Press.

ROE, F. J. C. and BÄR, A. (1985) Enzootic and epizootic adrenal medullary proliferative disease of rats: influence of dietary factors which affect calcium absorption, *Human Toxicology*, **4**, 27–52.

SANDUSKY, G. E., VAN PELT, C. S., TODD, G. C. and WIGHTMAN, K. (1988) An immunocytochemical study of pituitary adenomas and focal hyperplasia in old Sprague-Dawley and Fischer 344 rats, *Toxicologic Pathology*, **16**(3), 376–80.

SARTIN, J. L., CHAUDHURI, M., FARINA, S. and ADELMAN, R. C. (1986) Regulation of insulin secretion by glucose during aging, *Journal of Gerontology*, **41**, 30–5.

SASS, B., RABSTEIN. L. S., MADISON, R., NIMS, R. M., PETERS, R. L. and JONES, A. M. (1975) Incidence of spontaneous neoplasms in F344 rats throughout the natural life-span, *Journal of the National Cancer Institute*, **54**, 1449–56.

SILVERMAN, D. A. and ROSE, N. R. (1974) Neonatal thymectomy increases incidence of spontaneous and methylcholanthrene-enhanced thyroiditis in rats, *Science*, **184**, 162–3.

SOLLEVELD, H. A., HASSEMAN, J. R. and MCCONNELL, E. E. (1984) Natural history of body weight gain, survival and neoplasia in the F344 rat, *Journal of the National Cancer Institute*, **22**, 929–40.

SPENCER, A. J., ANDREU, M. and GREAVES, P. (1986) Neoplasia and hyperplasia of pancreatic endocrine tissue in the rat: an immunocytochemical study, *Veterinary Pathology*, **23**, 11–15.

STEFANEANU, L. and KOVACS, K. (1994) Changes in structure and function of the pituitary, in MOHR, U., DUNGWORTH, D. L. and CAPEN, C. C. (Eds), *Pathobiology of the Aging Rat*, pp. 173–91, Washington: ILSI Press.

STONE, J. P., HOLTZMAN, S. and SHELLBARGER, C. J. (1979) Neoplastic responses and correlated plasma prolactin levels in diethylstilboestrol-treated ACI and Sprague-Dawley rats, *Cancer Research*, **39**, 773–8.

STROMBERG, P. C., WILSON, F., CAPEN, C. C. (1983) Immunocytochemical demonstration of insulin in spontaneous pancreatic islet cell tumors of Fischer rats, *Veterinary Pathology*, **20**, 291–7.

SUZUKI, H., MOHR, U., KIMMERLE, G. (1979) Spontaneous endocrine tumors in Sprague-Dawley rats, *Journal of Cancer Research and Clinical Oncology*, **95**, 187–96.

TANG, F. (1985) Effect of sex and age on serum aldosterone and thyroid hormones in the laboratory rat, *Hormone and Metabolic Research*, **17**, 507–9.

TANNENBAUM, A., VESSELINOVITCH, S. D., MALTONI, C. and MITCHELL, S. D. (1962) Multipotential carcinogenicity of urethan in the Sprague-Dawley rat, *Cancer Research*, **22**, 1362–71.

THOMAS, G. A. and WILLIAMS, D. (1994) Changes in structure and function of the thyroid follicular cell, in MOHR, U., DUNGWORTH, D. L. and CAPEN, C. C. (Eds), *Pathobiology of the Aging Rat*, pp. 269–83, Washington: ILSI Press.

THURSTON, V. and WILLIAMS, E. D. (1982) Experimental induction of C-cell tumors in thyroid by increased content of vitamin D3, *Acta Endocrinologia*, **100**, 41–5.

TISCHLER, A. S., DeLELLIS, R. A., PERLMAN, R. L., ALLEN, J. M., COSTOPOULOS, D., LEE, Y. C., NUNNEMACHER, G., WOLFE, H. J. and BLOOM, S. R. (1985) Spontaneous proliferative lesions of the adrenal medulla in aging Long-Evans rats. Comparison of PC12 cells, small granule-containing cells and medullary hyperplasia, *Laboratory Investigation*, **53**, 486–98.

TREIP, C. S. (1983) The regression of oestradiol-induced pituitary tumours in the rat, *Journal of Pathology*, **141**, 29–40.

TUCKER, M. J. (1979) The effect of long term food restriction on tumours in rodents, *International Journal of Cancer*, **23**, 803–7.

UEBERBERG, H. and LUTZEN, L. (1979) The spontaneous rate of tumours in the laboratory rat: strain C4bb Thom (SPF), *Drug Research*, **29**, 1876–9.

VERSPOHL, E. J. and AMMON, H. P. T. (1983) Increased insulin binding to pancreatic islets of aged rats, *Endocrinology*, **112**, 2147–51.

WANG, S. Y., HALBAN, P. A. and ROWE, J. W. (1988) Effects of aging on insulin synthesis and secretion, *Journal of Clinical Investigation*, **81**, 176–84.

WARD, J. M. and REZNIK-SCHULLER, H. (1980) Morphological and histochemical characteristics of pigments in aging F344 rats, *Veterinary Pathology*, **17**, 678–85.

WIKLUND, J., WERTZ, N. and GORSKI, J. (1981) A comparison of oestrogenic effects on uterine and pituitary growth and prolactin synthesis in F344 and Holtzman rats, *Endocrinology*, **109**, 1700–7.

WILLIAMS, E. D. (1966) Histogenesis of medullary carcinoma of the thyroid, *Journal of Clinical Pathology*, **19**, 114–18.

WINKLER, H., APPS, D. K. and FISCHER-COLBRIE, R. (1986) The molecular function of adrenal chromaffin granules: established facts and unresolved topics, *Neuroscience*, **18**, 261–90.

WRIGHT, J., YATES, A., SHARMA, H. and THIBERT, P. (1985) Histopathological lesions in the pancreas of the BB Wistar rat as a function of age and duration of diabetes, *Journal of Comparative Pathology*, **95**, 7–14.

YANAGISAWA, M., HARA, Y., SATOH, K., TANIKAWA, T., SAKATSUME, Y., KATAYAMA, S., KAWAZU, S., ISHII, J. and KOMEDA, K. (1986) Spontaneous autoimmune thyroiditis in BioBreeding/Worcester (BB/W) rats, *Endocrinology* (Japan), **33**, 851–61.

YARRINGTON, J. T. and JOHNSTON, J. O'N. (1994) Aging in the adrenal cortex, in MOHR, U., DUNGWORTH, D. L. and CAPEN, C. C. (Eds), *Pathobiology of the Aging Rat*, pp. 227–44, Washington: ILSI Press.

ZAK, M., KOVACKS, K., McCOMB, D. J. and HEITZ, P. U. (1985) Amino-glutethamide-stimulated corticotrophs. An immunocytologic, ultrastructural and immunoelectron microscopic study of the rat adenohypophysis, *Virchows Archiv [B]*, **49**, 93–106.

12

The Nervous System

The nervous system is the primary control system of the body, with complex interactions with all organs and tissues. In the AP rat the brain, spinal cord and sciatic nerves have been examined routinely in standard toxicology studies. These tissues have been fixed by immersion but it is recognised that this method produces considerable artefact (Garman, 1990), and in toxicology studies any evidence of effects in the nervous system, whether clinical or morphological, warrant separate studies where perfusion fixation and special staining techniques can be employed.

12.1 Brain

In toxicology studies in the AP rat the meninges are cut at necropsy, and the whole brain removed and immersion fixed in buffered formalin for 24 hours. After this time it is removed from fixative for sampling. In all of the long term studies included in this database, the brains from all animals were examined, and samples for histological examination were taken by this author. In short term studies samples were taken by staff in the histology laboratory. The brain has had at least two sections in all toxicology studies, one passing through the cerebrum at the level of the infundibulum and the other through the middle of the cerebellum. In later years additional sections have been taken from the cerebrum and from the medulla oblongata.

The brain of the rat has several major differences from other species which may be of importance in toxicology. The rat brain, unlike the human, has no raised convolutions of the cerebral hemispheres (the gyri). In man the clefts (sulci) between the convolutions are prone to ischaemia and hypoxia – conditions which do not occur in the rat, where these anatomical features are missing. The substantia nigra of the rat has significantly less neuromelanin than primate species including man (D'Amato *et al.*, 1986). A sex difference in brain development in Long-Evans rats was reported by Diamond (1987): a thicker cortex is present on the right side in males and on the left in females. The differences appear to relate to different numbers of neurones and glial cells in the right and left sides.

12.1.1 Non-neoplastic Findings

Brain weights

Brain weights in AP rats at different time points up to 2 years are shown in Table 12.1. There appears to be a continuing growth of the brain until 2 years although there is no significant difference between 34 and 58 weeks. This would indicate that there is no significant atrophy up to 2 years nor any developmental neurological diseases (Walker *et al.*, 1989; Wright *et al.*, 1989). The weights in the AP rat are similar to those recorded by Krinke and Eisenbrandt (1994) in Wistar, SD and F344 rats.

Congenital anomalies

Congenital hydrocephalus is now a rare condition in the AP rat (<0.1 per cent) compared with a 5 per cent incidence when the strain was first used. The most severe hydrocephalus causes death within a few months. The incidence has declined in the AP strain by selective culling of the parents of affected offspring from the breeding colony. Clonic convulsions are rarely seen in the AP rat

Table 12.1 Brain weights in the AP rat

Age (weeks)	Number	Median weight of brain			
		Absolute weight (g)		Relative weight (% body weight)	
		Male	Female	Male	Female
12	10	1.92	1.71	0.56	0.58
34	20	2.25	1.98	0.36	0.62
58	25	2.20	1.97	0.30	0.51
110	80 (males)[a] 73 (females)	2.47	2.23	0.40	0.56

[a] This study included 150 rats/sex but the brain was not weighed in any animal which died before completion of two years or any animal which had macroscopic evidence of neoplasia.

although they are more common in Charles River Wistar rats, and in females (Nunn and Macpherson, 1995).

Inflammation

Meningitis and encephalitis are rare conditions in the AP rat. Both acute and chronic meningitis have been observed, with acute meningitis thought to be secondary to a large compressing pituitary adenoma. This is a very rare reaction to this type of tumour, which is a common occurrence in the AP rat. The few cases of mild diffuse chronic meningitis were incidental findings at routine histological examination and no obvious cause was apparent. The paucity of inflammatory lesions in the rat brain may be a reflection of their SPF status and of the specialised immune responses in the central nervous system.

Haemorrhage/infarcts/compression

Compression by pituitary tumours is the most common observation in the brain of the aged AP rat, where it is often accompanied by ventricular dilatation (Figure 106). Haemorrhage has only been observed as a consequence of compression; it is occasionally the ultimate cause of death in animals with very large tumours. Only a few focal infarcts have been seen, located in the cerebral cortex. Burek (1978) also reported a low incidence in BN rats over 30 months of age.

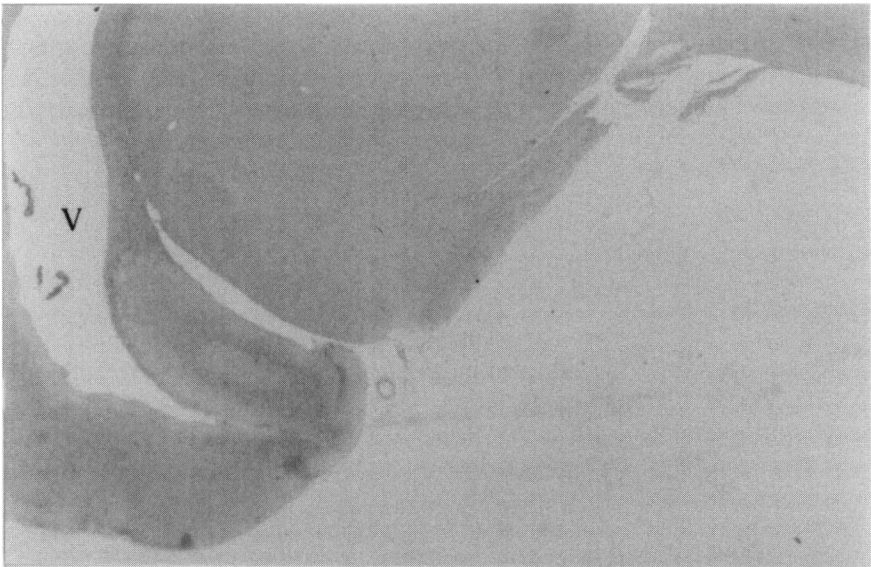

Figure 106 Compression of the cerebellum caused by a large pituitary tumour with dilatation of a lateral ventricle (V). ×8, H&E

Neuronal changes

The most common observation in the neurones of AP rats is lipofuscin pigmentation which occurs in at least some neurones of the cerebral cortex, spinal cord and Purkinje cells of all AP rats in 2 year studies. Similar pigmentation in the F344 has been described by Ward and Reznik-Schüller (1980) and Burek (1978). This lipofuscin accumulation is thought to be a consequence of peroxidation of membrane lipids by free radicals. Impairment of memory in rats has been correlated with increasing neuronal pigmentation (Kadar *et al.*, 1990). Neurones are as old as the brain in which they are located, as the cells do not divide nor are they replaced. This makes them susceptible to any material which can cross the blood–brain barrier, and heavy metals such as iron may accumulate in neurones.

Neuronal loss in older rats is still disputed (Flood and Coleman, 1988) but this may be due to different rates of loss in different areas of the brain. Rogers *et al.* (1984) reported a 25 per cent loss of Purkinje cells in old rats. Neuronal loss in other areas has been reported by Sabel and Stein (1981). Immuno-cytochemical staining for neurones has been a relatively recent development and can be used to demonstrate loss of neuronal cells. High levels of an isoenzyme of enolase are found in neurones (Schmechel, 1985), and although it is present in non-neuronal cells they do not express it to the same degree as neurones.

It has been postulated that aging in the hypothalamic neurones may have widespread effects in the endocrine system. In males, testosterone uptake by specific hypothalamic neurones is responsible for sexual differentiation (Sheridan *et al.*, 1974). Thus the decreasing testosterone levels in old males may damage the neurones. Similarly the cyclical changes in oestrogen levels in females may also damage neurones, and the high PRL levels in old female rats can damage the dopamine neurones in the hypothalamus. Changes in receptor status of these hormone-dependant neurones indicate impairment of the negative feedback control of hormones.

Mineralisation/gliosis

Small mineralised bodies, usually few in number, have been observed in the cerebellum of up to 5 per cent of male rats and 1 per cent of females. The bodies did not have any associated inflammation or evidence of any degenerative change. Gliosis is a rare condition (<0.1 per cent) and usually focal, and has been associated with small areas of necrosis and/or inflammation. Hypertrophy of astrocytes in old rats may represent a reaction to increased intracranial pressure from pituitary tumours (Krinke and Eisenbrandt, 1994).

Spongiform encephalopathy (status spongiosis)

Spongiform encephalopathy of the cerebral cortex has been observed in up to 2 per cent of AP rats over 2 years of age (Figures 107 and 108). The change is thought to be due to swollen astrocytes and resembles the histological changes produced by murine neurotropic retrovirus (Sharpe *et al.*, 1990). It is suggested that this spongiform change is the result of a metabolic defect in the astrocytes.

Demyelination

Vacuolation of the white matter was confirmed as demyelination by staining with Luxol fast blue and occurs in 2 to 3 per cent of AP rats in 2 year studies. Similar changes were reported by Burek (1978) in the BN and WAG strains.

12.1.2 *Neoplastic Changes*

Brain tumours are one of the more common types of tumour to occur in the AP rat. They are more common in males, and this has been reported in other strains (Fitzgerald *et al.*, 1974; Gopinath, 1986). There is also a preponderance of tumours in human males (Jones, 1986). Approximately half of the tumours have been derived from glial cells and half from the meninges and related tissues. The different histological types observed are shown in Table 12.2.

Astrocytomas are the most common glial tumour. They are densely cellular and poorly demarcated tumours (Figure 109) composed of uniform cells with round nuclei and the malignant tumours show necrosis with pseudopalisading (Figure 110). Perivascular cuffing by malignant cells, often in areas remote from

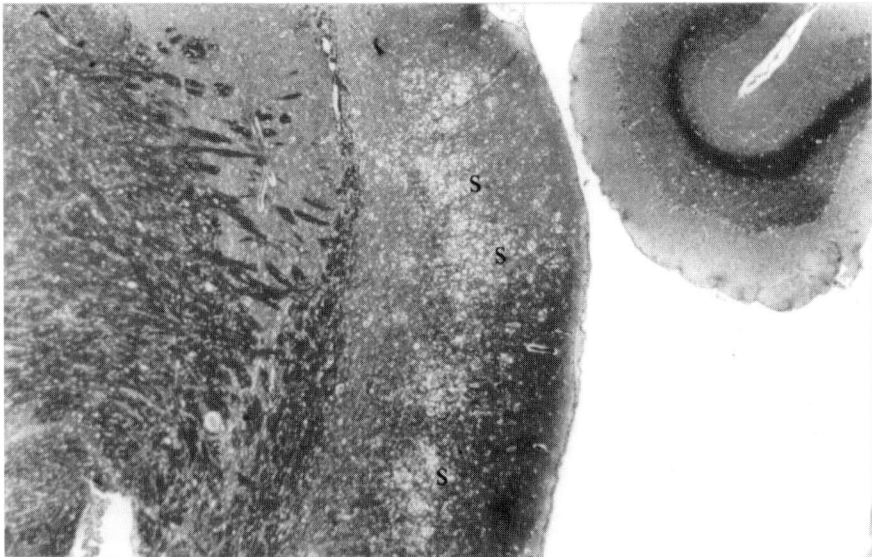

Figure 107   Status spongiosis (S) in the cerebral cortex of a male AP rat. ×8, H&E

Figure 108   Higher power view of status spongiosis. ×80, H&E

the tumour mass, is a common feature. All of the astrocytomas developed in the cerebrum, and infiltration was often widespread, but very few extended into the cerebellum. When unilateral the left side was more frequently affected than the right. The distinction between early gliomas and focal gliosis is difficult in the rat. The criteria used in the AP rat has been the presence of cellular atypia and an infiltration for the early glioma, while gliosis is characterised by the small size of

Table 12.2   Histological types of brain tumour observed in the AP rat

Histological type	Highest % incidence observed or total number[a]	
Astrocytoma	5%	
Meningioma/granular cell tumours	5%	
Oligodendrogliomas		3
Mixed glioma		2
Medulloblastoma		4[b]
Ependymoma		3
Malignant fibrohistiocytic sarcoma		1
Pinealoma		1
Adrenal ganglioneuroma[c]		6

[a] Incidence of tumours in control groups from 24 long term studies.
[b] All medulloblastomas were found in animals <2 months old.
[c] Adrenal ganglioneuromas are discussed in the chapter on the endocrine system.

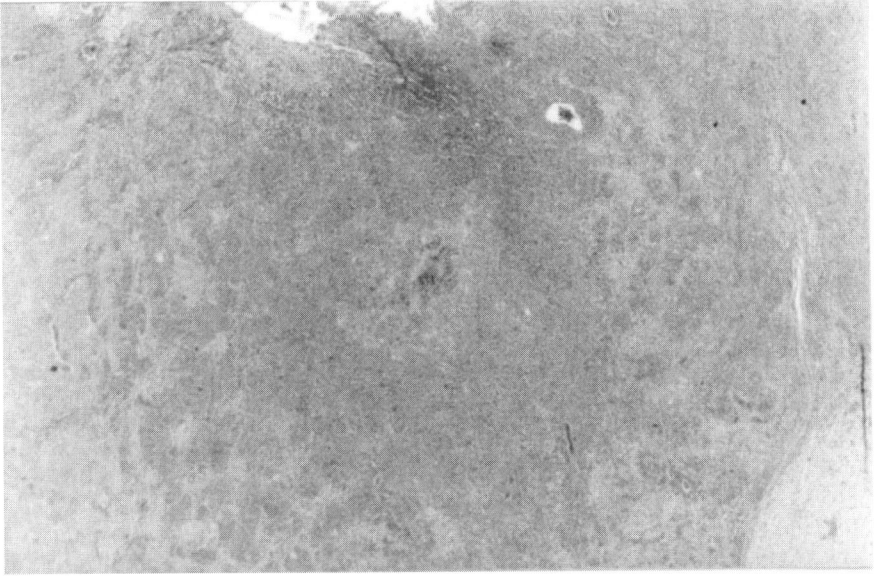

Figure 109 Diffusely infiltrating astrocytoma in a male AP rat aged 22 months. ×8, H&E

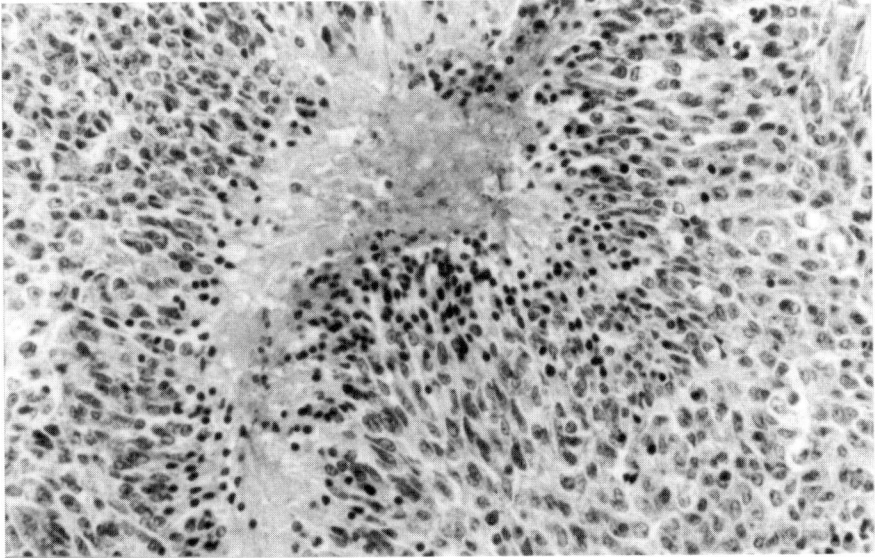

Figure 110 Malignant astrocytoma: pseudopalisading and necrosis of astrocytes. ×128, H&E

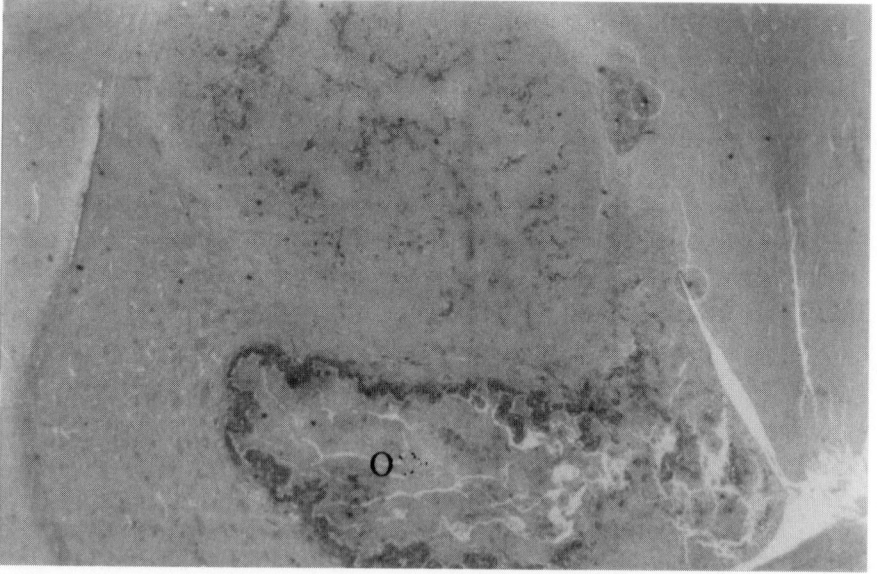

Figure 111   Oligodendroglioma (O) in a male AP rat. ×8, H&E

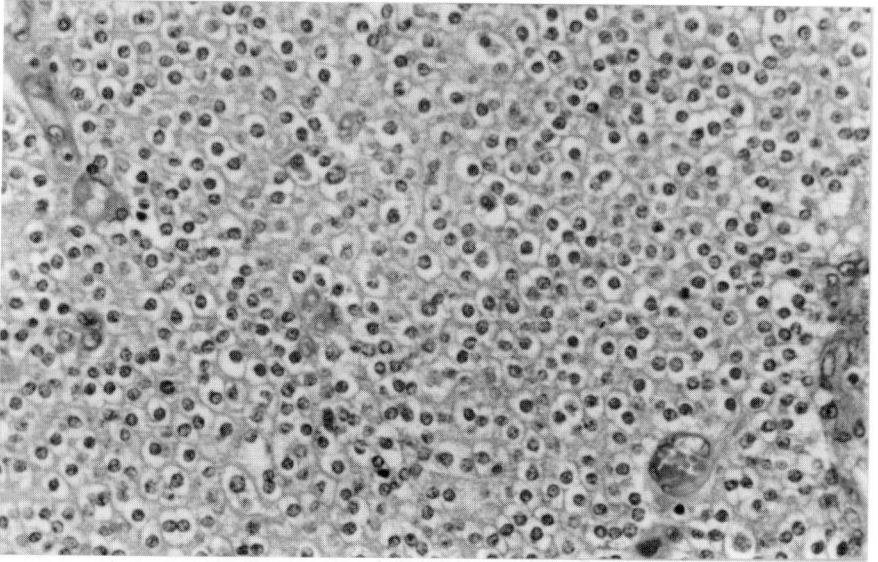

Figure 112   Oligodendroglioma showing a dense mass of uniform cells in honeycomb arrangement. ×80, PAS

the area involved, generally clear demarcation, and the presence of fibrosis, haemorrhage or inflammation.

Oligodendrogliomas were clearly demarcated, often with necrosis (Figure 111), densely cellular (Figure 112), with uniform cells showing clear cytoplasm, round or ovoid nuclei and a characteristic honeycomb appearance (Figure 113). The mixed gliomas included both astrocytes (the majority cell) and areas of oligodendrocytes.

The two ependymomas were found in the fourth ventricle and were densely cellular tumours, with round or oval nuclei, indistinct cell boundaries and rosette formation (Figure 114). Only one pinealoma has been found and this was composed of large pale cells and smaller dark cells in a perivascular distribution. The medulloblastomas were all found in rats less than 2 months of age and were the cause of death in each animal. They were highly infiltrative with indistinct boundaries. The cells were pleomorphic with elongated nuclei and indistinct cell boundaries and there was a high mitotic rate. The one fibrohistiocytic sarcoma was thought to have arisen in the meninges and showed the histological appearance of this tumour at other sites.

Tumours of the meningeal and associated tissues were as common as the gliomas, but were distributed throughout the brain with no preference for any particular anatomical site. Seventy per cent were in the cerebrum and 30 per cent in the cerebellum. The majority of these tumours (90 per cent) were meningiomas. They were clearly demarcated (Figure 115) and frequently showed marked compression (Figure 116). The cells showed abundant

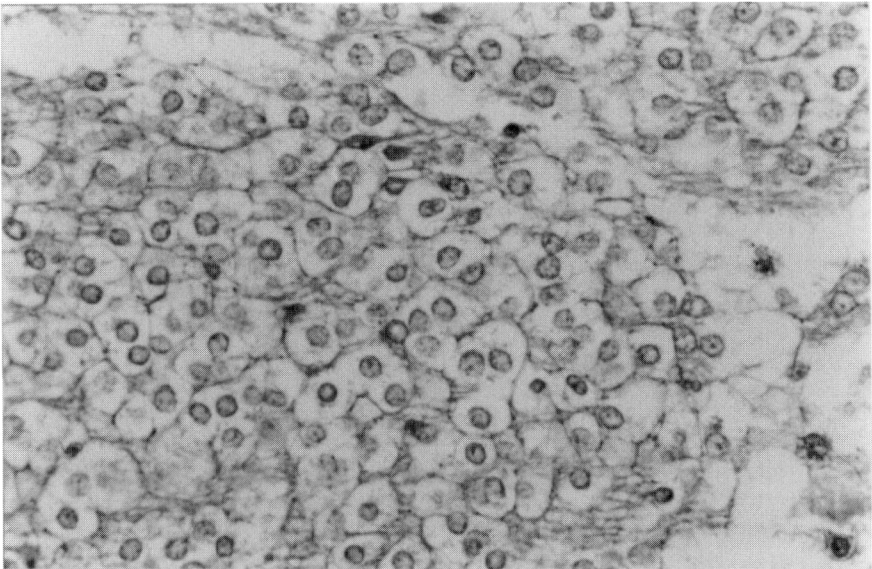

Figure 113 Oligodendroglioma showing cells with clear cytoplasm and small round nuclei. ×128, PAS

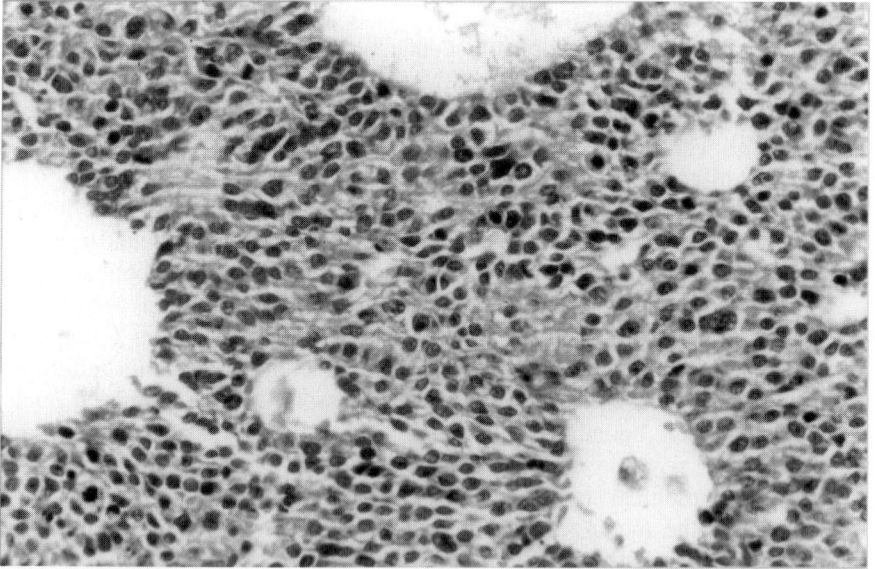

Figure 114 Ependymoma in a male AP rat: a densely cellular tumour with rosette formation. ×80, H&E

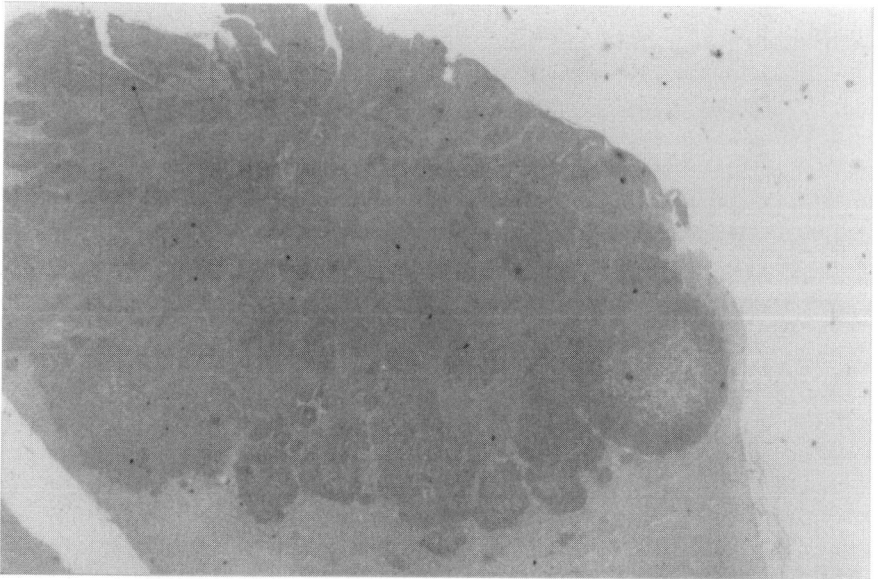

Figure 115 Large meningioma in a male AP rat showing the clear margin of the tumour. ×8, H&E

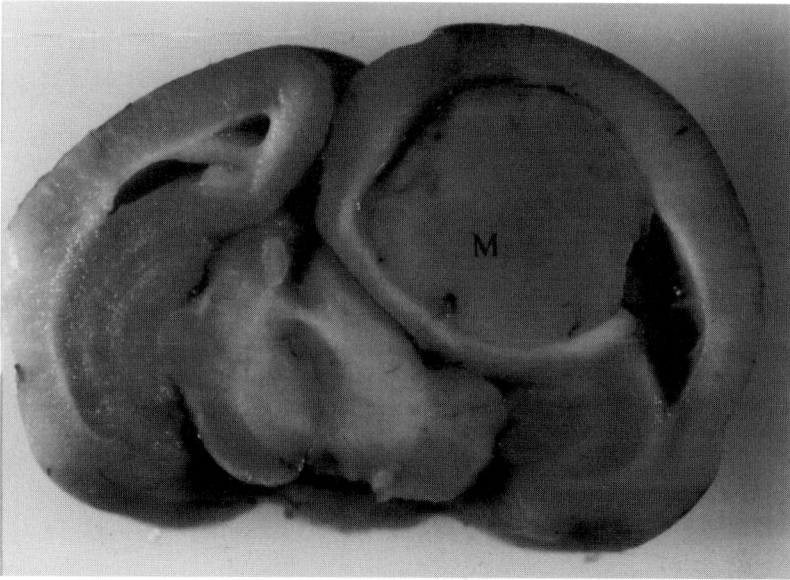

Figure 116   Brain from an AP rat sectioned through a meningioma (M) showing marked compression

eosinophilic cytoplasm and a variable fibrous stroma. About 5 per cent of the meningiomas were spindle-cell tumours arranged in interweaving bundles. Several had numerous calcified bodies (Figure 117). The granular cell neoplasms were of very large cells with abundant granular cytoplasm, which was periodic acid-Schiff positive, and large vesicular or small elongated nuclei (Figure 118).

Most of the meningeal tumours (70 per cent) were incidental findings and were not considered to be an important cause of death. Glial tumours, by contrast, occurred at an earlier age, with several occurring before 12 months of age. The clinical signs associated with brain tumours in AP rats are shown in Table 12.3. Some workers have reported the absence of clinical signs associated in animals with brain tumours (Gopinath, 1986), and in the AP rat the large majority of animals with meningeal tumours did not show any clinical signs. There was a range of signs in animals with gliomas, with a significant proportion showing nasal haemorrhage, presumably due to increased intracranial pressure as the tumours had not spread outside the brain.

Secondary tumours of the brain include leukaemias and lymphomas which were chiefly confined to meningeal infiltration and only rarely extended into the brain parenchyma.

The incidence of brain tumours in the AP rat is shown in Table 12.4. There has been a significant increase in the incidence of brain tumours with time, and the incidence of gliomas is higher in males while meningeal tumours have a similar incidence in both sexes. The reason for the increase in tumours of the

227

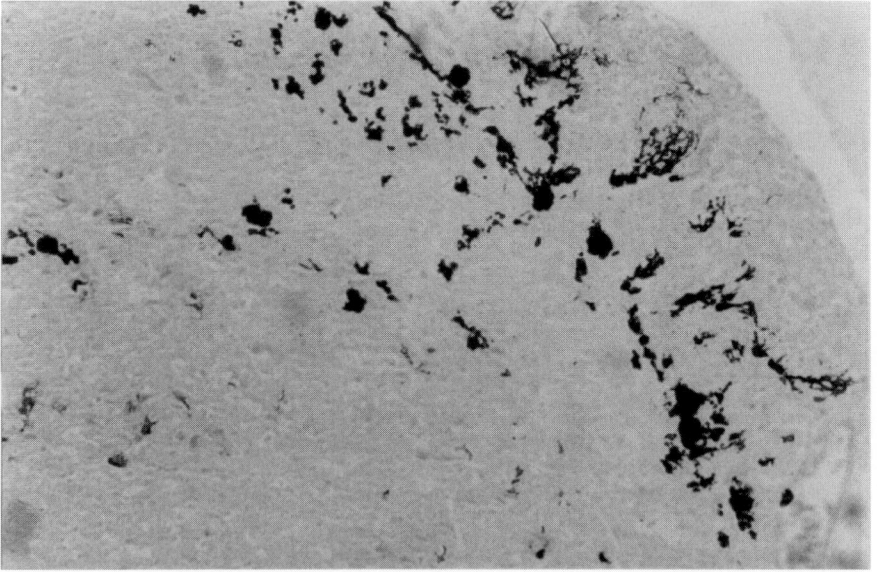

Figure 117 Numerous black-stained calcified bodies in a male AP rat. ×32, Von Kossa

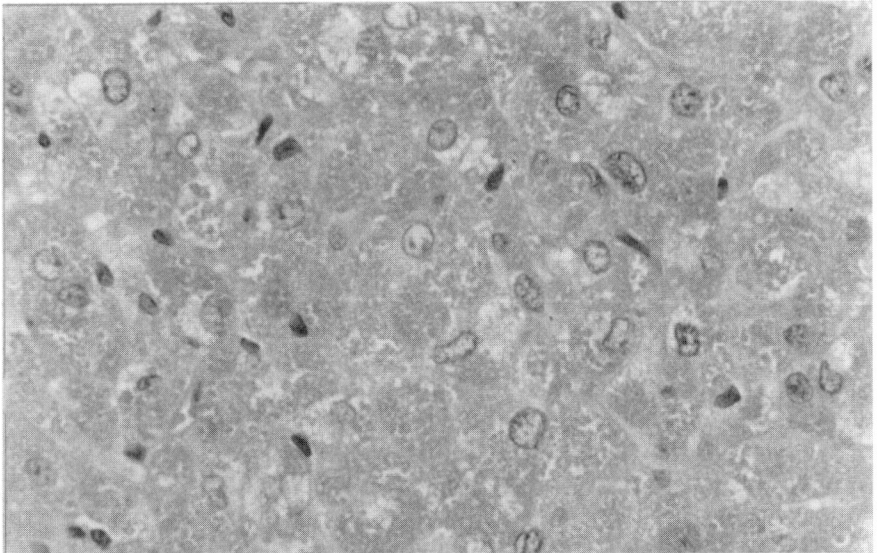

Figure 118 Granular cell meningeal tumour: large cells with an abundant granular cytoplasm and round pale nuclei. ×129, H&E

Table 12.3 Type and incidence of clinical signs associated with brain tumours in the AP rat

Clinical sign	% Incidence of clinical sign	
	Rats with gliomas	Rats with meningiomas
No clinical sign	20	70
Weight loss	21	9
Ataxia	5	8
Paralysis	5	6
Nasal haemorrhage	10	1
Increased aggression	6	5

Table 12.4 Incidence of brain tumours in the AP rat

Year study completed	Incidence of brain tumours[a]	
	Male	Female
1963 (Lifespan)	3/194 (1.5%)	2/288 (0.7%)
1971	1/25 (4.0%)	0/25
1971	2/50 (4.0%)	1/50 (2.0%)
1973	0/60	2/60 (3.0%)
1974	5/65 (7.6%)	0/65
1975	3/65 (4.6%)	2/65 (3.0%)
1975	4/65 (6.1%)	4/65 (6.1%)
1975	4/50 (8.0%)	1/50 (2.0%)
1976	2/34 (5.8%)	2/34 (2.0%)
1978	7/50 (14.0%)	4/50 (8.0%)
1982	23/300 (7.6%)	8/300 (2.6%)
1986	6/100 (6.0%)	4/100 (4.0%)
1990	11/100 (11.0%)	3/100 (3.0%)

[a] Incidence of brain tumours from a database of 8800 AP rats (including 2800 males and 2500 females in 2 year studies) used in studies between 1960 and 1992. All studies except the life-span study (51 months) were of 24 months duration. Nine studies between 1963 and 1971 had no animal with a brain tumour.

brain is not known. It does not appear to be due to the increase in body size and food intake which has occurred in the strain, since food restriction has no effect on brain tumours (Tucker, 1979). Variability of brain tumour incidence has been observed by Swenberg (1986) and Solleveld *et al.* (1984). The differences in incidence may be due to differences in the examination of the brain. In the AP rat 10 per cent of tumours were located at post-fixation examination of the brain and would have been missed if standard samples had been taken. This may account for the general low incidences (<6 per cent) reported for many strains such as the F344 (Solleveld *et al.*, 1984; Haseman *et al.*, 1990), the SD rat (Gopinath, 1986; Krinke *et al.*, 1985) and the Osborne-Mendel rat (Goodman *et al.*, 1980).

12.2 Spinal Cord

In toxicology studies in the AP rat the spinal cord is taken, *in situ*, in the lumbar vertebrae. After immersion fixation, longitudinal and transverse sections are taken for histological examination. In general the spinal cord shows similar pathological changes to the brain.

12.2.1 *Non-neoplastic Changes*

Inflammation/cysts/pigmentation

No significant inflammation has been seen in the spinal cord but small perivascular mononuclear cell infiltrates are seen occasionally. A few animals have also shown small fluid-filled cysts in the spinal cord (Figure 119). Lipofuscin is also seen in spinal neurones in older animals.

Degenerative myelinopathy

This is a very common condition in older AP rats. It is characterised by the

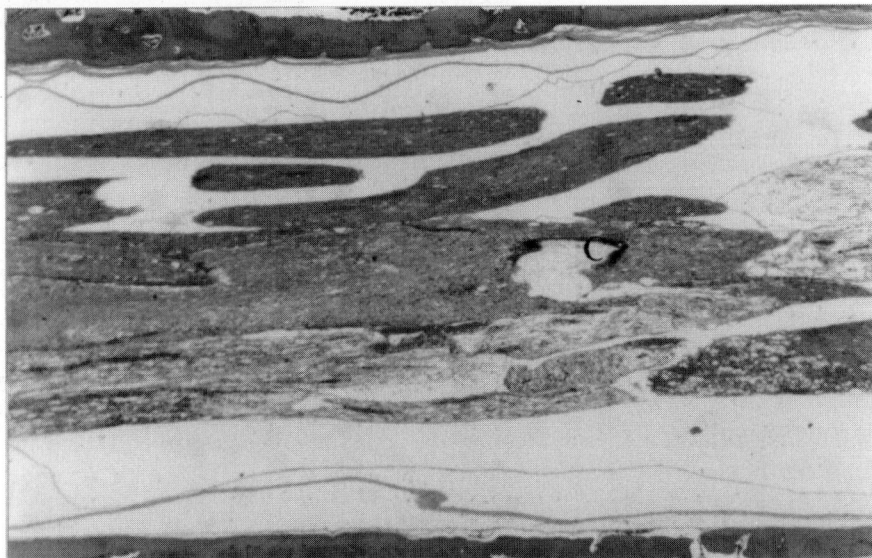

Figure 119 Fluid-filled cyst (C) in the spinal cord of a female AP rat. ×32, H&E

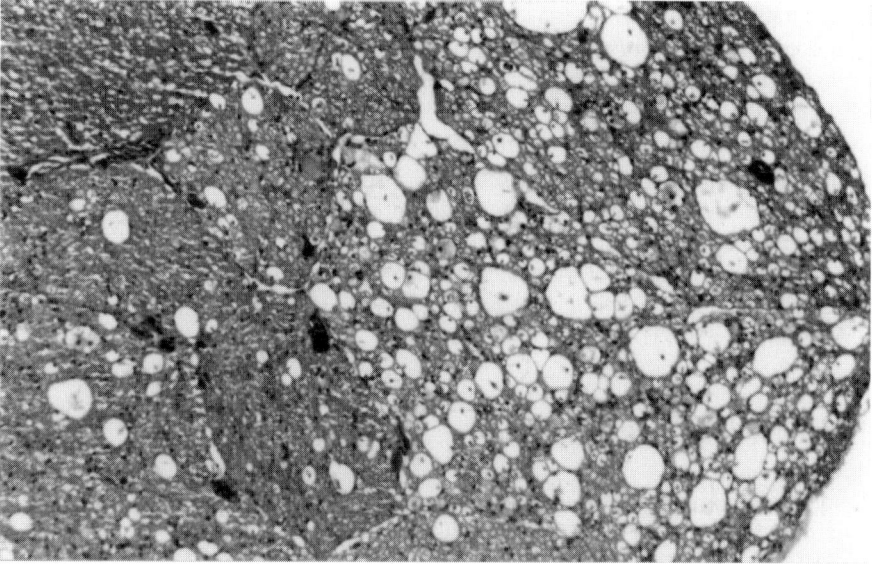

Figure 120 Cystic myelinopathy in the spinal cord of a male AP rat aged 26 months. ×80, H&E

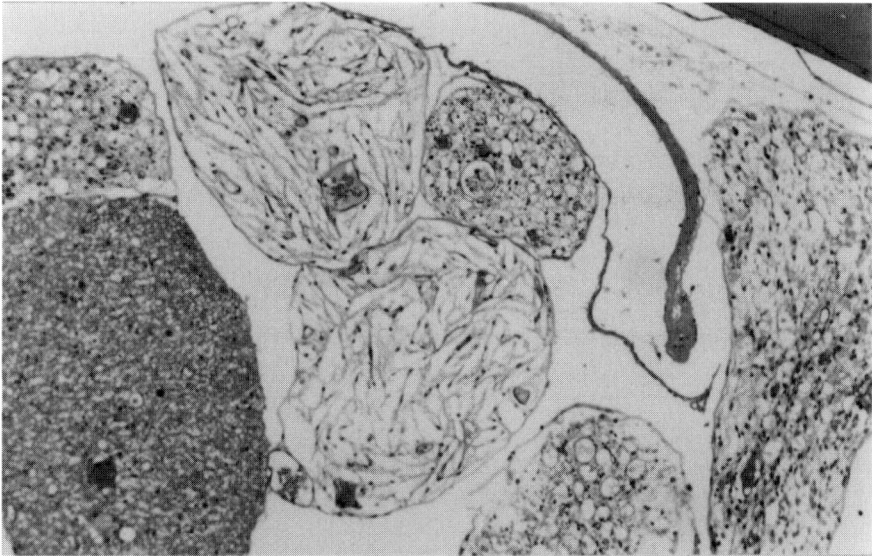

Figure 121 Degenerative myelinopathy in spinal nerve roots showing 'cholesterol' clefts. ×80, H&E

degeneration of the myelinated fibres of the white matter. There is swelling and degeneration of myelin sheaths and axons. In the most severe cases the myelin has a cystic appearance (Figure 120). The condition frequently involves the spinal nerve roots where clefts, possibly cholesterol clefts, may be numerous (Figure 121).

This condition can affect up to 70 per cent of AP rats at two years with a similar incidence in males and females, although only approximately 10 per cent will show the severe lesions which are accompanied by muscle atrophy and paralysis. Minimal changes may be observed in animals of 14 months of age, but the incidence does not increase significantly until the animals reach 18 months of age. This lesion has been reported in other strains (Burek, 1978; Krinke *et al.*, 1981).

12.2.2 Neoplastic Changes

Two astrocytomas of the lumbar spinal cord have been found in AP rats and a meningioma of the cervical cord.

12.3 Peripheral Nerves

The left sciatic nerve has been examined in routine toxicology studies with the AP rat.

12.3.1 Non-neoplastic Changes

Spinal radiculoneuropathy

The only important change seen in the sciatic nerve of the AP rat is the degenerative spinal radiculoneuropathy which is much more common in males than females. This is a progressive, segmental demyelination commencing with swelling of the myelin sheaths. As the sheaths degenerate and fragment, macrophages infiltrate the endoneurium to ingest the myelin and a granulomatous reaction may occur (Greaves and Faccini, 1984). A detailed description of the microscopic and ultrastructural changes has been given by King (1994). In the most severe condition there may be inflammation, oedema and pockets of fibrosis (Figure 122) and small foci of calcification. The incidence of this radiculoneuropathy can reach 46 per cent in males with 10 per cent showing a severe lesion; in females the overall incidence in 2 year studies can reach up to 10 per cent and the severity is only minimal to mild. Whether this reflects a temporal difference in development or a real difference in susceptibility is not known, as sciatic nerves were not examined in the life-span study.

232

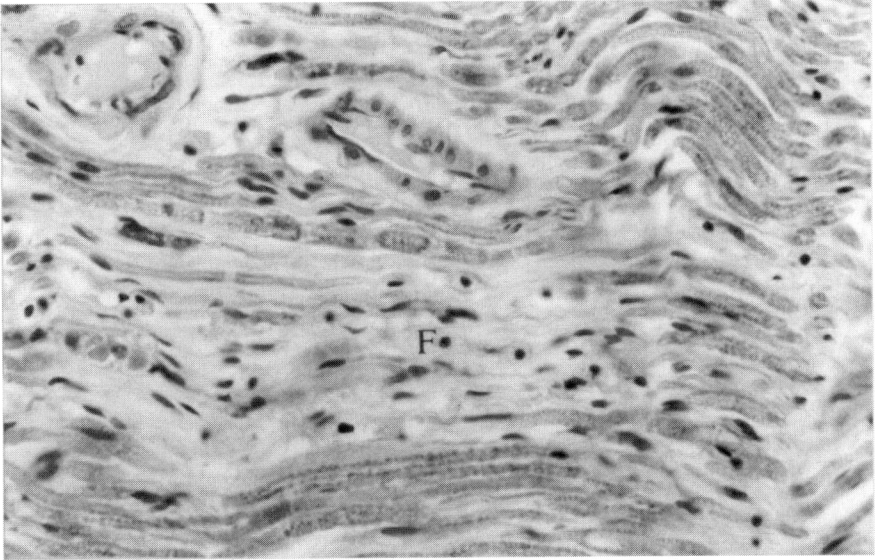

Figure 122   Area of fibrosis (F) in the sciatic nerve of a 20 month old male AP rat. ×128, H&E

Few other nerves in the AP rat have been studied in sufficient numbers to give an accurate assessment of the incidence of the disease, except for the brachial plexus, where an incidence of 20 per cent (24/93) was seen in males and 24 per cent (20/83) in females in one 2 year study. In both sexes only a mild degeneration was observed. More extensive nerve damage in males, compared with females, is also reported in other strains (van Steenis and Kroes, 1971; Cotard-Bartley *et al.*, 1981). Changes have been reported in many different peripheral nerves although the severity, onset and progression vary considerably. Cotard-Bartley *et al.* (1981) considered that the sciatic and tibial nerves were affected most frequently. Changes have been reported in the plantar and tibial nerves (Thomas *et al.*, 1980) and tail nerves (King and Thomas, 1983). The cause of this common condition in the rat is not known. Cotard-Bartley *et al.* (1981) thought that rats kept in wire cages could develop the disease because wire cages prevent coprophagy and produce a deficiency of vitamins B12 and B1 which are necessary for nerve function. Vitamin E deficiency is also associated with axonal degeneration (Southam *et al.*, 1991). Reducing food intake delays the onset of muscle atrophy but not radiculoneuropathy (Berg *et al.*, 1962). Pressure damage to the plantar nerves may occur in obese rats kept in wire cages (King, 1994), but this cannot be the cause in other nerves. It is most likely that the changes in most nerves are a function of aging, possibly due to changes in blood flow (Kihara *et al.*, 1991) or loss of adrenergic control of the microcirculation (Koistinaho *et al.*, 1990).

12.3.2 Neoplastic Changes

A single Schwannoma has been seen in a male AP rat arising in the sciatic nerve. Spontaneous peripheral nerve tumours are rare in all strains of rat (Gough *et al.*, 1986).

12.4 References

BERG, B. N., WOLF, A. and SIMMS, H. S. (1962) Nutrition and longevity in the rat IV. Food restriction and the radiculoneuropathy of aging rats, *Journal of Nutrition*, **77**, 439–42.

BUREK, J. D. (1978) *Pathology of Aging Rats*, West Palm Beach, Florida: CRC Press.

COTARD-BARTLEY, M. P., SECCHI, J., GLOMOT, R. and CAVANAGH, J. B. (1981) Spontaneous degenerative lesions of peripheral nerves in aging rats, *Veterinary Pathology*, **18**, 110–13.

D'AMATO, R. J., LIPMAN, Z. P. and SNYDER, S. H. (1986) Selectivity of the Parkinsonian neurotoxin MPTP: Toxic metabolite MPP$^+$ to neuromelanin, *Science*, **231**, 987–9.

DIAMOND, M. C. (1987) Sex differences in the rat forebrain, *Brain Research Reviews*, **12**, 235–40.

FITZGERALD, J. E., SCHARDEIN, J. L. and KURTZ, S. M. (1974) Spontaneous tumours of the nervous system in albino rats, *Journal of the National Cancer Institute*, **52**, 265–73.

FLOOD, D. G. and COLEMAN, P. D. (1988) Neuron numbers and sizes in aging brain: comparisons of human, monkey and rodent data, *Neurobiology of Aging*, **9**, 453–63.

GARMAN, R. H. (1990) Artefacts in routinely immersion-fixed nervous tissue, *Toxicologic Pathology*, **18**, 149–53.

GOODMAN, D. G., WARD, J. M., SQUIRE, R. A., CHU, K. C. and LINHART, M. S. (1980) Neoplastic and non-neoplastic lesions in aging Osborne-Mendel rats, *Toxicology and Applied Pharmacology*, **55**, 433–47.

GOPINATH, C. (1986) Spontaneous brain tumours in Sprague-Dawley rats, *Food and Cosmetic Toxicology*, **24**, 433–47.

GOUGH, A. W., HANNA, W., BARSOUM, N. J., MOORE, J. and STURGESS, J. M. (1986) Morphologic and immunohistochemical features of two spontaneous peripheral nerve tumours in Wistar rats, *Veterinary Pathology*, **23**, 63–73.

GREAVES, P. and FACCINI, J. M. (1984), in *Rat Histopathology*, pp. 216–17, Amsterdam: Elsevier.

HASEMAN, J. K., ARNOLD, J. and EUSTIS, S. L. (1990) Tumor incidences in Fischer 344 rats: NTP historical data, in BOORMAN, G. A., EUSTIS, S. L. and ELWELL, M. R. (Eds). *Pathology of the Fischer Rat*, pp. 555–64, San Diego: Academic Press.

JONES, R. D. (1986) Epidemiology of brain tumours in man and their relationship with chemical agents, *Food and Chemical Toxicology*, **24**, 99–103.

KADAR, T., SILBERMAN, M., BRADEIS, R. and LEVY, A. (1990) Age-related structural changes in the rat hippocampus: correlation with working memory deficiency, *Brain Research*, **512**, 113–20.

KIHARA, M., NICKANDER, K. K. and LOW, P. A. (1991) The effect of aging on endoneural blood flow, hyperemic response and oxygen-free radicals in rat sciatic nerve, *Brain Research*, **562**, 1–5.

KING, R. H. M. (1994) Changes in the peripheral nervous system, in MOHR, U., DUNGWORTH, D. L. and CAPEN, C. C. (Eds), *Pathobiology of the Aging Rat*, Vol. 2, pp. 35–53, Washington: ILSI Press.

KING, R. H. M. and THOMAS, P. K. (1983) Distal axonal degeneration in ageing rats, *Neuropathology and Applied Neurobiology*, **9**, 73–4.

KOISTINAHO, J., WADHWANI, K. C. and RAPOPORT, S. I. (1990) Adrenergic innervation of the tibial and vagus nerves in rats of different ages. *Mechanisms of Ageing and Development*, **52**, 195–205.

KRINKE, G. J. and EISENBRANDT, D. L. (1994) Non-neoplastic changes in the brain, in MOHR, U., DUNGWORTH, D. L. and CAPEN, C. C. (Eds), *Pathobiology of the Aging Rat*, Vol. 2, pp. 1–19, Washington: ILSI Press.

KRINKE, G., SUTER, J. and HESS, R. (1981) Radicular myelinopathy in aging rats, *Veterinary Pathology*, **8**, 335–41.

KRINKE, G., NAYLOR, D. C., SCHMID, S., FRÖHLICH, E. and SCHNEIDER, K. (1985) The incidence of naturally occurring primary brain tumours in the laboratory rat, *Journal of Comparative Pathology*, **95**, 175–92.

NUNN, G. and MACPHERSON, A. (1995) Spontaneous convulsions in Charles River Wistar rats, *Laboratory Animals*, **29**, 50–3.

ROGERS, J., ZORNETZER, S. F., BLOOM, F. E. and MERVIS, R. E. (1984) Senescent microstructural changes in rat cerebellum, *Brain Research*, **292**, 23–32.

SABEL, B. A. and STEIN, D. G. (1981) Extensive loss of subcortical neurons in the aging rat brain, *Experimental Neurology*, **73**, 507–16.

SCHMECHEL, D. E. (1985) γ-subunit of the glycolytic enzyme enolase: Non-specific or neuron specific? *Laboratory Investigation*, **52**, 239–42.

SHARPE, A. H., HUNTER, J. J., CHASSIER, P. and JAENISCH, R. (1990) Role of abortive retroviral infection of neurons in spongiform CNS degeneration, *Nature*, **346**, 181–3.

SHERIDAN, P. J., SAR, M. and STUMPF, W. E. (1974) Autoradiographic localisation of H3-testosterone or its metabolites in the neo-natal rat brain, *American Journal of Anatomy*, 140, 589–93.

SOLLEVELD, H. A., HASEMAN, J. K. and MCCONNELL, E. E. (1984) Natural history of body weight gain, survival and neoplasia in the F344 rat, *Journal of the National Cancer Institute*, **72**, 929–40.

SOUTHAM, E., THOMAS, P. K. and KING, R. H. M. (1991) Experimental vitamin E deficiency in rats. Morphological and functional evidence of abnormal axonal transport secondary to free radical damage, *Brain*, **114**, 915–36.

SWENBERG, J. A. (1986) Brain tumours – problems and perspectives, *Food and Chemical Toxicology*, **24**, 155–8.

THOMAS, P. K., KING, R. H. M. and SHARMA, A. K. (1980) Changes with age in the peripheral nerves of the rat, *Acta Neuropathology*, **54**, 83–7.

TUCKER, M. J. (1979) The effect of long-term food restriction on tumours in rodents, *International Journal of Cancer*, **23**, 803–7.

VAN STEENIS, G. and KROES, R. (1971) Changes in the nervous system and musculature of old rats, *Veterinary Pathology*, **8**, 320–32.

WALKER, A. F., GUERRIERO, F. J., TOSCANO, T. V. and WEIDENMAR, C. A. (1989) Relative cerebellar weight: a potential indicator of developmental neurotoxicity, *Neurotoxicology and Teratology*, **11**, 251–5.

WARD, J. M. and REZNIK-SCHULLER, H. (1980) Morphological and histochemical characteristics of pigments in aging F344 rats, *Veterinary Pathology*, **17**, 678–85.

WRIGHT, J. A., BRICKELL, L. and FOSTER, J. (1989) The neuroteratogenicity of procarbazine hydrochloride in the rat: histopathological and immunocytochemical aspects, *Journal of Comparative Pathology*, **101**, 421–7.

13

Special Sense Organs and Associated Tissues

13.1 Eyes

Both eyes and Harderian glands have been examined in toxicology studies with the AP rat. In the first 20 years the eyes were immersion fixed in Zenker's acetic, but latterly Davidson's fluid has been used. Histological sections are taken through the eye at the level of the optic nerve. During the course of short term toxicity studies the eyes of rats are examined, at intervals, by ophthalmoscope, after dilating the pupils with Mydriacyl (Alcon Laboratories, Watford, UK). In two 2-year studies routine examination of the eye is confined to examination of the anterior segment by ophthalmoscope immediately prior to necropsy. Spontaneous lesions in the eye of the rat have been comprehensively reviewed by Taradach and Greaves (1984) and Yoshitomo and Boorman (1990).

13.1.1 Non-neoplastic findings

Congenital abnormalities

Anophthalmia and microphthalmia are rare conditions. An animal with two lenses in one eye has been observed, and persistent hyaloid artery is seen occasionally.

Inflammation

Inflammation of the cornea (keratitis) and conjunctiva occurs principally as a result of infection with the sialodacryoadenitis (SDA) corona virus in the AP rat.

Apart from studies which have shown SDA infection, keratitis is uncommon in the AP rat, with incidence levels of 1–2 per cent in 2 year studies. In one 2 year study, which was infected with SDA virus in the first year, clinical examination during the infection period showed an 80 per cent incidence of corneal lesions, but histological examination after terminal necropsy revealed few sequelae. The histological appearance of keratitis is of a flattened corneal epithelium with desquamated surface cells and varying degrees of inflammation and fibrosis. Ulceration of the cornea has only been seen in panophthalmitis. Iritis is even less common (<1 per cent) and, when observed, has been secondary to inflammation elsewhere in the anterior segment.

Corneal dystrophy (mineralisation)

Mineral deposits in the cornea occurred in <1 per cent of AP rats in most studies, except for the one with the SDA infection where the incidence was 10 per cent. The granular deposits were either in the basement membrane or below it in the centre of the cornea. This change has been reported in several strains of rat (Taradach and Greaves, 1984; Bellhorn *et al.*, 1988; Losco and Troup, 1988) with incidence levels ranging between 10 per cent in some F344 strains to 35 per cent in the SD rat, and up to 66 per cent in some other Wistar strains. The aetiology is not yet known. A similar condition in humans (calcific band keratopathy) has been associated with alterations in calcium metabolism. This has not yet been proved in the rat (Weisse, 1994). Vascularisation of the cornea can be induced by various dietary factors such as diets deficient in vitamin A or zinc (Carter-Dawson *et al.*, 1980; Leurre-Dupree, 1986). In most cases vascularisation follows inflammation, and Klintworth and Burger (1983) consider that leucocytes and activated macrophages are involved in the development of this condition.

An increase in thickness of Descemet's membrane has been observed in old (>18 months) AP rats, and this was reported in another Wistar strain by Weisse *et al.* (1974).

Haemorrhage/pigmentation

Haemorrhage into the anterior chamber has been observed as a result of trauma (Figure 123) and pigmentation of the cornea may occur after haemorrhage.

Retinal atrophy

Atrophy of the retina in albino rats, such as the AP rat, is thought to be due to age (senile atrophy) or exposure to high ambient light intensity. The atrophy in the AP rat is focal in the early stage, and progresses to a severe diffuse loss of cells from both the inner and outer nuclear layers (Figures 124–7). Similar changes have been recorded for ageing rats of many strains, including the SD rat (Lin and Essner, 1987), F344 (Coleman *et al.*, 1977; Lee *et al.*, 1990), and

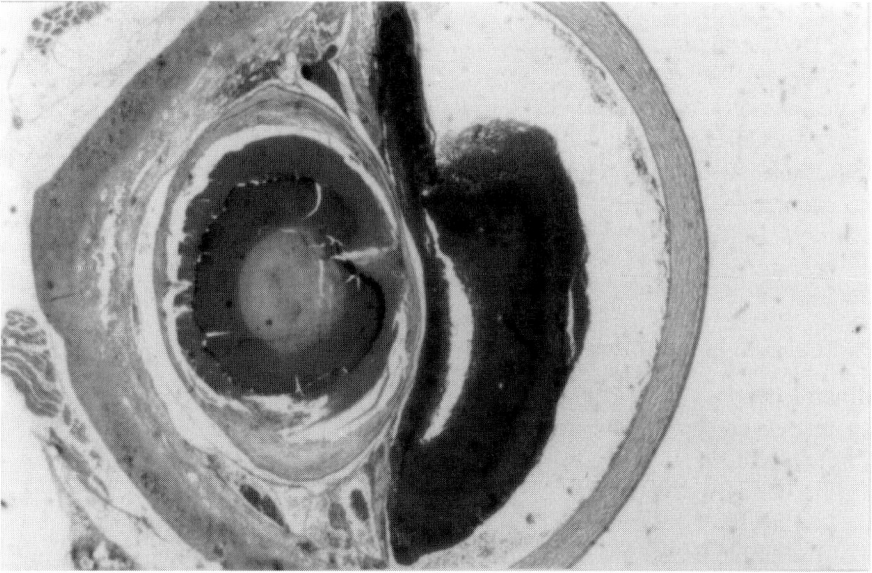

Figure 123   Haemorrhage into the anterior chamber of the eye of a male AP rat. ×8, H&E

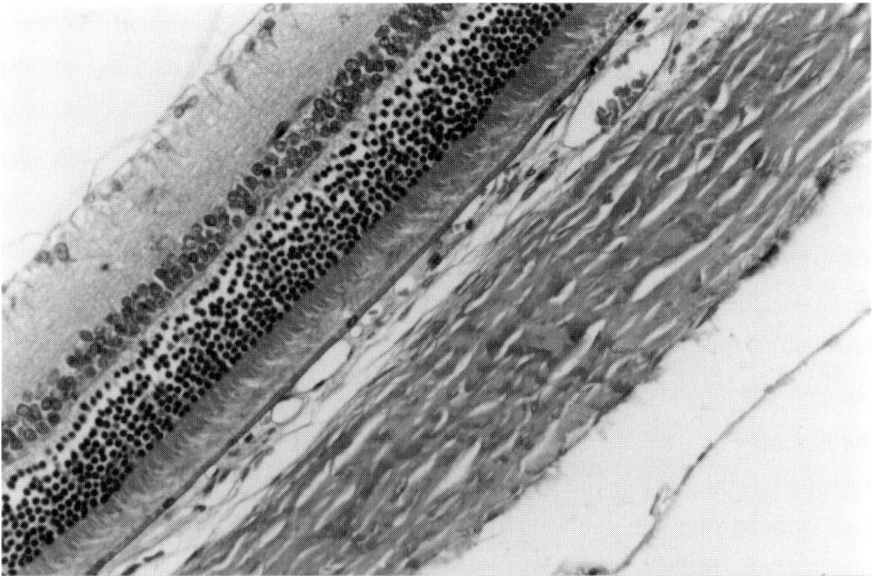

Figure 124   Normal retina in a male AP rat. ×128, H&E

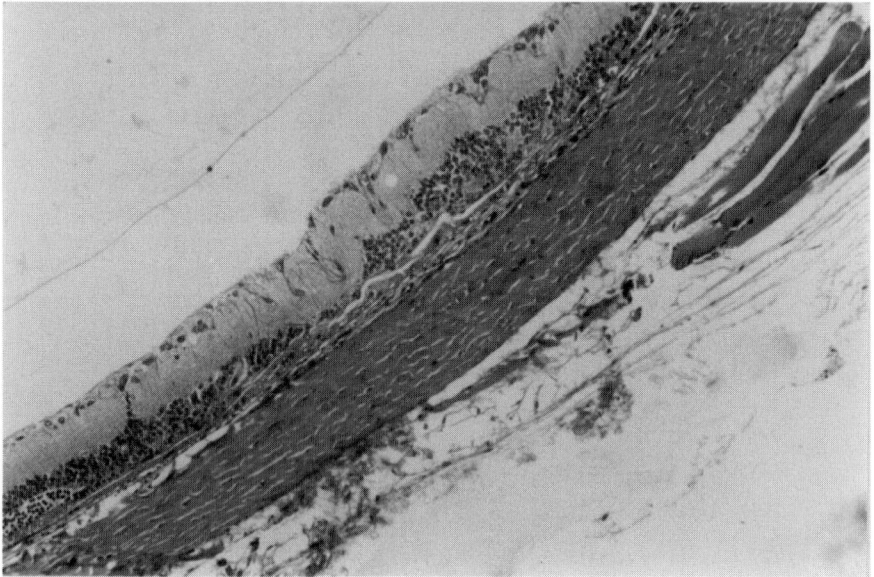

Figure 125   Mild focal retinal atrophy in an AP rat with loss of the outer nuclear layer. ×80, H&E

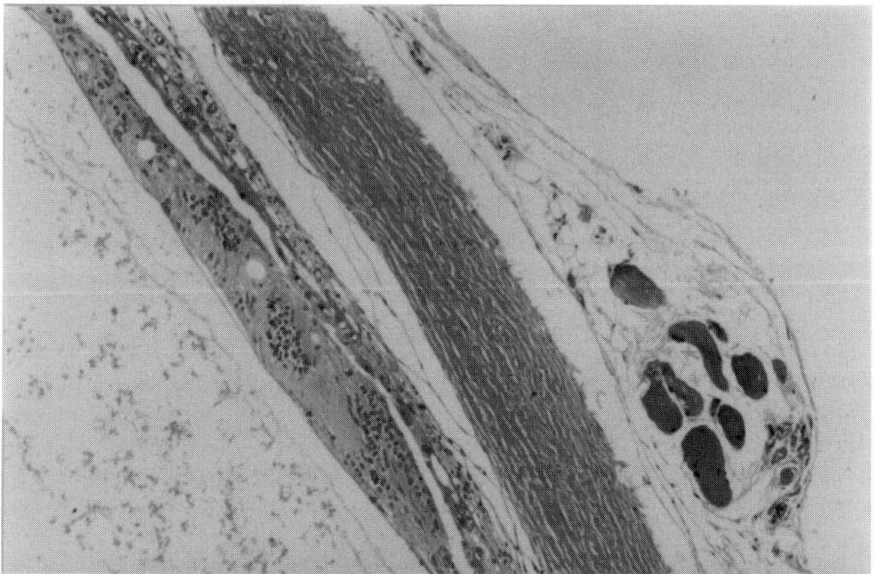

Figure 126   Severe diffuse retinal atrophy in an AP rat with disorganisation and loss of both nuclear layers. ×80, H&E

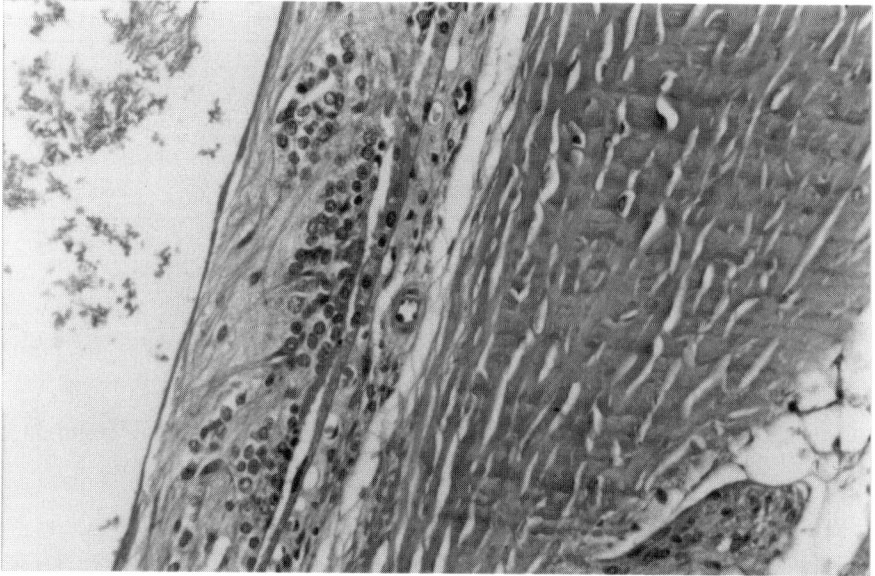

Figure 127   Higher power view of severe retinal atrophy. ×128, H&E

other Wistar strains (Mawdsley-Thomas, 1968). The incidence in the AP rat in 2 year studies can reach 10 per cent and there is no difference in the incidence in males and females. In one study in the early 1960s, where animals were exposed to natural daylight, the incidence of retinal atrophy in the AP rats which were housed nearest to the south-facing windows was five-fold higher than in the animals which were on the opposite side of the rack where the cages faced inwards towards a wall. Photoxic retinal degeneration has been reported by Noell and Albrecht (1971) and Grignolo *et al.* (1969). The RCS and Hunter rat strains have an inherited retinal degeneration which commences in very young animals (Dowling and Sidman, 1962; Yates *et al.*, 1974).

Lens

The transparency of the lens depends on the solubility of the proteins contained within it. There are three types of soluble proteins termed crystallines, and some insoluble proteins located in the centre of the lens, in the older fibres. About 50 per cent of the lens is composed of water, and the integrity of the tissue depends on diffusion of nutrients, chiefly from the aqueous humour. The chief nutrient is glucose which undergoes glycolysis to form lactic acid; this then diffuses back out of the lens into the aqueous humour. The rate of entry of glucose into the lens and the rate of glycolysis within it are controlled by hexokinase. Any factor which disturbs the equilibrium, and this includes dietary and metabolic factors, can change the nature of the crystalline proteins and cause an opacity or cataract. In the AP rat two types of cataract have been

241

Table 13.1   Incidence of cataracts at week 86 of a 2 year study in control AP rats

| | Incidence of cataracts | | | |
| | Crystalline | | Crescentic | |
Eye	Males	Females	Males	Females
Right	4/300	7/300	4/300	17/300
Left	5/300	2/300	1/300	20/300
Bilateral	2/300	3/300	0/300	5/300
Total	11/300 (3.6%)	14/300 (4.6%)	5/300 (1.6%)	42/300 (14%)

identified: crystalline and crescentic cataracts. Both types are more common in females than males; a typical incidence in a 2 year study is shown in Table 13.1.

The ultimate appearance of a severe cataract is similar, whether crescentic or crystalline, with disruption, swelling and ballooning of lens fibres (Figure 128). Cataracts may be focal or diffuse and the crystalline type can arise in the anterior or posterior cortex. The crescentic cataracts are unilateral, crescentic in shape and originate in the dorsal equator. These cataracts are fully formed by six months of age and do not progress with time. Some of the cataracts, diagnosed by ophthalmology, disappeared completely. The crescentic cataract in the AP rat has been described in detail by Lazenby *et al.* (1993).

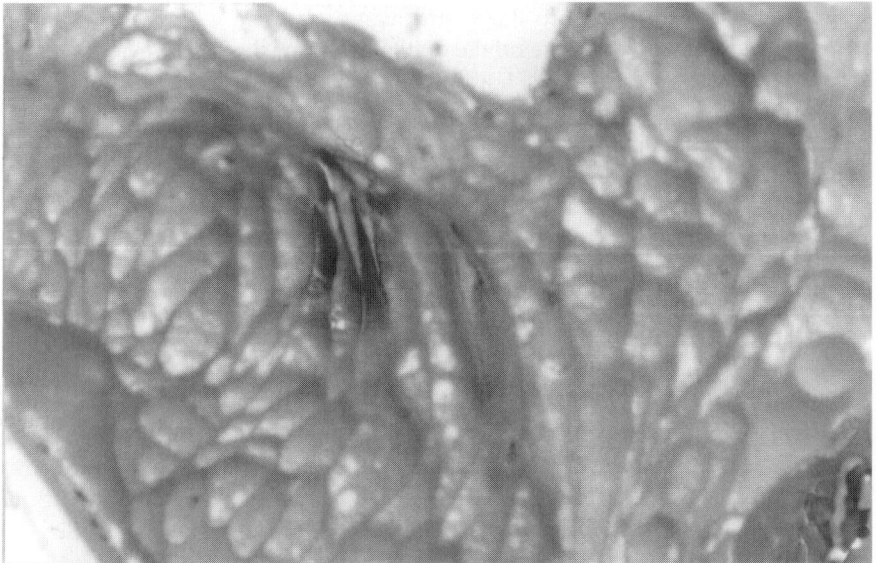

Figure 128   Severe crystalline cataract in a male AP rat; swelling and ballooning of lens fibres. ×128, H&E

13.1.2 Neoplastic Changes

Primary intraocular tumours are rare in all strains of rat (Heywood, 1975; Taradach and Greaves, 1984) and have been cited in detail by Weisse (1994). In the AP rat three have been identified: a squamous carcinoma of the conjunctiva, a fibroma in the choroid and an amelanotic melanoma of the retina (diagnosed by electron microscopy).

13.2 Harderian Glands

13.2.1 Non-neoplastic Changes

The harderian glands are associated with the inner canthus of the eye, while the inferior lacrimal glands, which include the infraorbital and exorbital glands, are associated with the outer canthus. Only the harderian glands have been examined routinely in toxicology studies in the AP rat, although the exorbital gland is frequently present in samples of the salivary glands. The harderian gland secretes lipid substances and porphyrin pigment which is more extensively produced by females (Sakai, 1981). The secretion is affected by vitamin B deficiency (Eida *et al.*, 1975).

Inflammation

Sialodacryoadenitis in the rat, caused by rat corona virus infection, causes a red tinged ocular and nasal discharge (chromodacryorrhea) with a necrotising dacryoadenitis, glandular dilatation, stromal inflammation and increased porphyrin (Figures 129 and 130). The incidence is low in the AP rat. Small foci of chronic inflammation occur sporadically in AP rats of all ages, and acinar dilatation can occur in the absence of inflammation. Necrotising adenitis can arise as a consequence of orbital blood sampling (McGee and Maronpot, 1979).

13.2.2 Neoplastic Changes

Only three tumours of the harderian gland have been observed in AP rats: two adenomas and one adenocarcinoma. Exophthalmus was only evident with the large malignant tumour. The benign tumours were small, well-differentiated tumours of clearly defined acini of cells with the characteristic foamy cell cytoplasm; a few mitoses were present and the tumours had a fine capsule. The adenocarcinoma was poorly differentiated and infiltrative, with acini often showing a vacuolated papillary epithelium, or solid acini with a dense eosinophilic cytoplasm (Figure 131). Few spontaneous tumours have been recorded in other strains (Goodman *et al.*, 1979; Rothwell and Everitt, 1986).

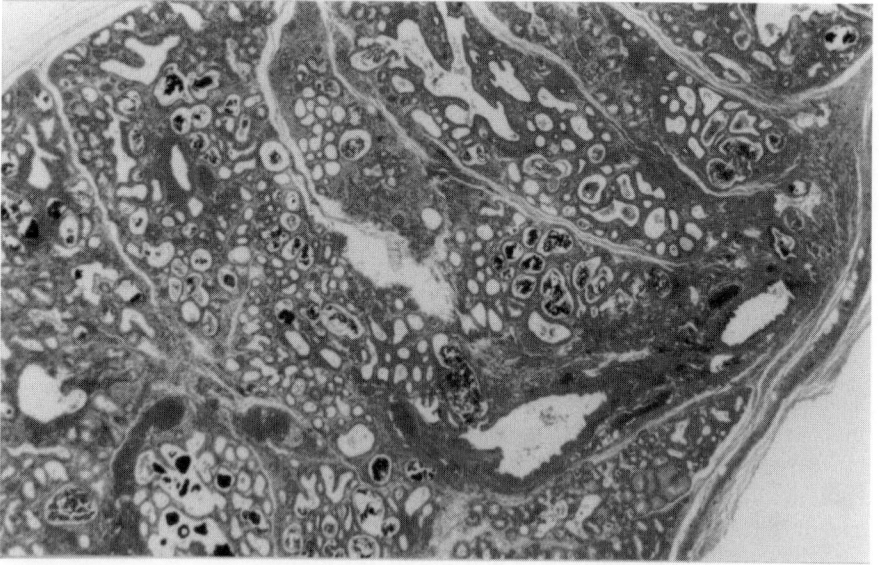

Figure 129 Dacryoadenitis showing glandular dilatation, stromal inflammation and increased porphyrin. ×32, H&E

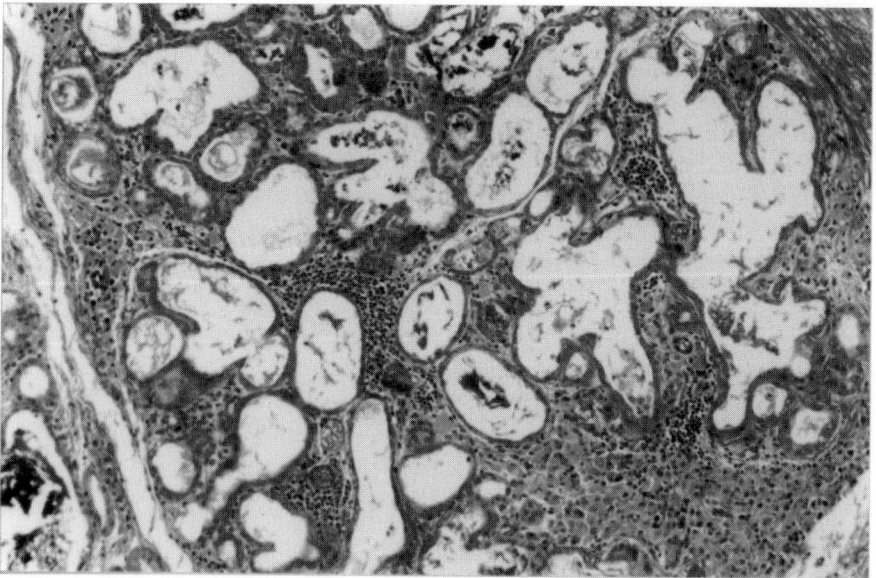

Figure 130 Higher power view of dacryoadenitis. ×80, H&E

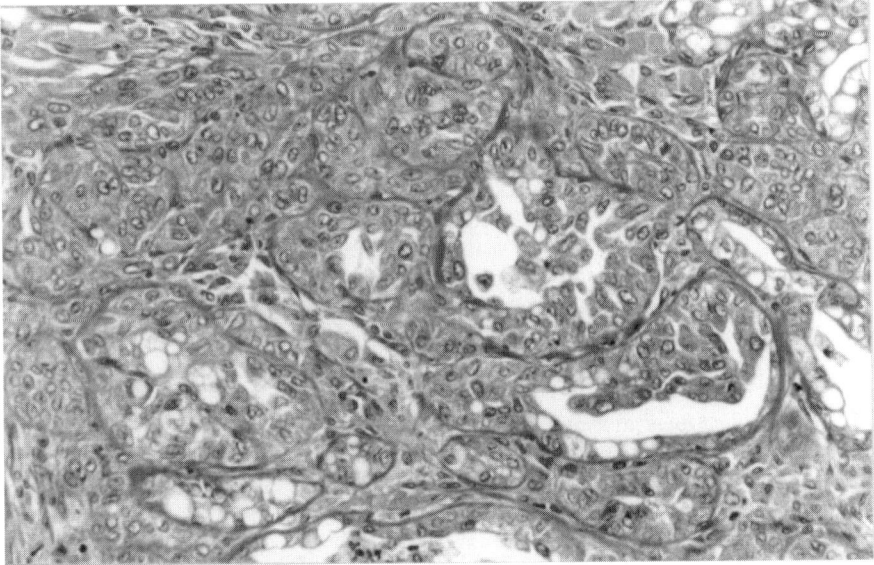

Figure 131 Adenocarcinoma of the harderian gland showing acini with a papillary, vacuolated epithelium. ×80, H&E

13.3 Exorbital gland

Cytomegaly of the acinar epithelium and inflammation have been observed incidentally in sections of the extraorbital glands, but no neoplastic lesions have been seen.

13.4 Ear

The ear has not been examined routinely in the AP rat. A few animals have exhibited the characteristic twirling movements of otitis media, and sections of the head at the level of the ear have shown inflammation or abscesses in the middle ear. Auricular chondritis has also been seen in a few animals. The pinna has been greatly swollen due to a chronic inflammation of the cartilage and surrounding subcuticular tissue. Small foci of regenerative cartilage were also present. No tumours have been observed in the ear, but tumours of the auditory gland (Zymbal's gland) are described in Chapter 2.

13.5 References

BELLHORN, R. W., KORTE, G. E. and ABRUTYN, D. (1988) Spontaneous corneal degeneration in the rat, *Laboratory Animal Science*, **38**, 46–50.

CARTER-DAWSON, L., TANKA, M., KUWABARA, T. and BIERI, J. G. (1980) Early corneal changes in vitamin A deficient rats, *Experimental Eye Research*, **30**, 261–8.

COLEMAN, G. L., BARTHOLD, S. W., OSBALDISTON, G. W., FOSTER, S. J. and JONAS, A. M. (1977) Pathological changes during aging in barrier-reared Fischer 344 male rats, *Journal of Gerontology*, **32**, 258–78.

DOWLING, J. E. and SIDMAN, R. L. (1962) Inherited retinal dystrophy in the rat, *Journal of Cell Biology*, **14**, 73–109.

EIDA, K., KUBOTA, N., NISHIGAKI, T. and KIKUTANI, M. (1975) Harderian gland. V. Effect of dietary pantothenic acid deficiency on porphyrin biosynthesis in harderian gland of rats, *Chemical and Pharmacology Bulletin* (Tokyo), **23**, 1–4.

GOODMAN, D. G., WARD, J. M., SQUIRE, R. A., CHU, K. C. and LINHART, M. S. (1979) Neoplastic and non-neoplastic lesions in aging F344 rats, *Toxicology and Applied Pharmacology*, **48**, 237–48.

GRIGNOLO, A., ORZALESI, N., CASTELLAZZO, R. and VITONE, P. (1969) Retinal damage by visible light in albino rats: an electron microscope study, *Ophthalmology*, **157**, 43–59.

HEYWOOD, R. (1975) Glaucoma in the rat, *British Veterinary Journal*, **131**, 213–21.

KLINTWORTH, G. K. and BURGER, P. C. (1983) Neovascularisation of the cornea: Current concepts of its pathogenesis, *International Ophthalmology Clinics*, **23**, 27–39.

LAZENBY, C. M., WESTWOOD, F. R. and GREAVES, P. (1993) Crescentic cataracts in Alderley Park rats, *Veterinary Pathology*, **30**, 70–4.

LEE, E. W., RENDER, J. A. and GARNER, C. D. (1990) Unilateral degeneration of retina and optic nerve in Fischer 344 rats, *Veterinary Pathology*, **27**, 439–44.

LEURRE-DUPREE, A. E. (1986) Vascularization of the rat cornea after prolonged zinc deficiency, *Anatomical Record*, **216**, 27–32.

LIN, W. L. and ESSNER, E. (1987) An electron microscopic study of retinal degeneration in Sprague-Dawley rats, *Laboratory Animal Science*, **38**, 703–10.

LOSCO, P. E. and TROUP, C. M. (1988) Corneal dystrophy in Fischer 344 rats, *Laboratory Animal Science*, **38**, 702–10.

MAWDSLEY-THOMAS, L. E. (1968) Retinal atrophy in the Wistar rat, *Excerpta Medica International Congress Series*, **181**, 164–74.

MCGEE, M. A. and MARONPOT, R. R. (1979) Harderian gland dacryoadenitis in rats resulting from orbital bleeding, *Laboratory Animal Science*, **29**, 639–41.

NOELL, W. K. and ALBRECHT, R. (1971) Irreversible effects of visible light on the retina: the role of vitamin A, *Science*, **172**, 76–9.

ROTHWELL, T. L. W. and EVERITT, A. V. (1986) Exophthalmos in ageing rats with Harderian gland disease, *Laboratory Animals*, **20**, 97–100.

SAKAI, T. (1981) The mammalian harderian gland. Morphology, biochemistry and physiology, *Archives Histology Japan*, **44**, 299–333.

TARADACH, C. and GREAVES, P. (1984) Spontaneous eye lesions in laboratory animals: incidence in relation to age, *CRC Critical Reviews in Toxicology*, **12**, 121–47.

WEISSE, I. (1994) Aging and ocular changes, in MOHR, U., DUNGWORTH, D. L. and CAPEN, C. C. (Eds), *Pathobiology of the Aging Rat*, pp. 77–9, Washington: ILSI Press.

WEISSE, I., STOTZER, II. and SEITTZ, R. (1974) Age and light dependant changes in the rat eye, *Virchows Archiv [A]*, **362**, 145–56.

YATES, C. M., DEWAR, A. J. and WILSON, H. (1974) Histological and biochemical studies on the retina of a new strain of dystrophic rat, *Experimental Eye Research*, **18**, 119–33.

YOSHITOMO, K. and BOORMAN, G. A. (1990) Eye and associated glands, in BOORMAN, G. A., EUSTIS, S. L. and ELWELL, M. R. (Eds), *Pathology of the Fischer Rat*, pp. 239–59, San Diego and New York: Academic Press.

Index

Adrenal gland 193–201
  accessory nodules 196
  adrenal weights 194
  anatomy 194
  atrophy 198
  capsule 196
  cellular alteration 198
  cortical lipidosis 197
  ectopic bone 196
  functional changes 195
  haemocysts 197
  hyperplasia 198–9
  inflammation 196
  mineralisation 196
  necrosis 196
  neoplasia 199–201
  pigmentation 198
adrenaline 195, 198
adrenocorticotrophic hormone (ACTH) 184,
    195
albumin 12, 14, 87
Alderley Park (AP) rat 1–21
aldosterone 195
alkaline phosphatase 12, 13
alveolar hyperplasia 116
angioma
  skin 27–30
  liver 71
  lymph node 135–6
angiosarcoma 27
anophthalmia 2, 13
antidiuretic hormone 184
apoptosis 66
arteritis 105–6
  mesentry 105
  muscle 38

ovary 150
spleen 128
testes 165, 167
tongue 51
uterus 155
atrophy
  adrenal gland 198
  bone marrow 123–4
  ovary 147–8
  pancreas 72–3
  prostate 175
  seminal vesicle 173
  spleen 127
  testes 165–6
astrocytoma 221–3
auricular chondritis 245

bile duct proliferation 66–7
bone 40–6
  chondromucoid degeneration 41–2
  fracture 41
  hyperostosis 43
  neoplasia 43–4
  osteoporosis 41–3
bone marrow 9, 123–6
  atrophy 123–4
  hyperplasia 124
  myelofibrosis 124
  neoplasia 124–6
brain 217–230
  brain weights 218
  compression 219
  demyelination 221
  encephalitis 219
  gliosis 220
  haemorrhage 219

brain (*continued*)
   infarct 219
   meningitis 219
   mineralisation 220
   neoplasia 221–30
   neuronal changes 220
   pigmentation 187
   spongiform encephalopathy 220–1
bromocriptine 106
bronchial associated lymph tissue (BALT) 116
bronchioalvoelar adenocarcinoma 119
bronchopneumonia 113

caging 4
calcitonin 203
calcium homeostasis 43, 204
castration cells 185
cataracts 241–2
catecholamines 195
c-cells 203–4, 206–7
ceroid 150, 198
cervix 156
cholesterol 12, 14, 195
cholesterol clefts 231, 232
choriocarcinoma 157
chromodacryorrhea 84, 186, 243
chronic progressive glomerulonephropathy
   (CPGN) 82–7, 199
chronic progressive cardiomyopathy 98–101
cirrhosis 69
clinical chemistry 12–14
coagula 92
coagulating gland 177
corneal dystrophy 238
corona virus 51, 243
corpora amylacea 176
corpora lutea 146–8
corticosterone 195
cystic degeneration 133–4
cystic ovarian bursa 149
cystitis 92
cysts
   adrenal 197
   lymph node 133
   ovary 149
   pituitary 185
   spleen 127
   thyroid 205

decidual reaction 156
Descemet's membrane 238
degenerative myelinopathy 230–1
demyelination 221
diet 4–5, 34, 63–4, 84–5, 148–9, 190
dopamine agonist 85, 106
dysgerminoma 151, 154

ear 245
   auricular chondritis 245
   otitis media 245

encephalitis 219
endocardial fibrosis 103–4
endometrial hyperplasia 155
endometrial polyps 157–8
endometritis 155
endoplasmic reticulum 64
β–endorphin 184
enteritis 59
ependymoma 222, 225
epididymides 171–2
   inflammation 171
   microcystic degeneration 171
   neoplasia 172
erythrpoietin 123
exorbital gland 245
extramedullary haematopoiesis 127
eyes 237–43
   cataract 241–3
   congenital anomalies 237
   corneal dystrophy 238
   haemorrhage 238
   keratitis 237–8
   lens 241
   neoplasia 243
   pigmentation 238
   retinal atrophy 238–41

fallopian tube 156–7
   salpingitis 156
fat necrosis 25
follicular stimulating hormone (FSH) 15–16,
   146, 148, 165, 184
food restriction: effects on
   arteritis 105
   bile duct proliferation 66–7
      calcitonin 204
      kidneys 85–6
      Leydig cells 168
      muscle 38
      ovaries 148–9
      pituitary tumours 190
      prostate 176
      radiculoneuropathy 233
      salivary gland weights 53

gastritis 56–7
gingivitis 47
gliosis 220
α_{2u}–globulin 14, 87
glucose 12, 14
gonadotrophin stimulating hormones 146
granulosa cell tumours 151
growth hormone 184
growth rate 5–7

haematology 9–11
haemocysts adrenal 197
hair emboli 106, 117
harderian glands 243–5
   chromodacryorrhea 243

dacryoadenitis 243
  neoplasia 244–5
hibernoma 27–8
heart 97–104
  cardiac weights 97–8
  cartilagenous metaplasia 101
  endocardial fibrosis 103–4
  hypertrophy 103
  mineralisation 101–2
  myocarditis 98–101
  neoplasia 104
  osseous metaplasia 101
  pericarditis 98
  thrombosis 103
  valvular degeneration 102
histiocytosis 132–4
hydrocephalus 16, 218
hydrometra 155
hydronephrosis 87

incisors 47
imperforate vagina 159
infarct 65, 89
insulin 37, 191, 193
insulitis 192
interstitial cells
  ovary 147
  testes 168
intestines 59–62
  congenital anomalies 59
  ectopic pancreas 59
  enteritis 59–60
  mineralisation 59
  nematodes 59
  neoplasia 60–2
  Peyers patches 59
  iodine 203, 207
islets of Langerhans 191–3
  alpha cells 191
  beta cells 191, 193
  delta cells 191
  glucagon 191
  inflammation 192
  neoplasia 193
  pigmentation 192
  polypeptide cells 191

joints 40

keratitis 238
kidneys 81–91
  congenital anomalies 81
  chronic progressive glomerulonephropathy
    (CPGN) 82–7
  hydronephrosis 87
  infarct 89
  kidney weights 81–2
  neoplasia 89–91
  nephrocalcinosis 88
  oncocytes 89

pigmentation 89
protein droplets 87
pyelonephritis 87–8

leukaemia 9, 11
  monocytic 129
  myeloid 124–5
  lymphatic 125
Leydig cell tumours 168–71
lipidosis 197–8
liver 62–72
  altered foci 69
  apoptosis 70
  bile duct proliferation 66–70
  cirrhosis 69
  clear cell change 63–4
  hepatic phosphorylase kinase deficiency 63
  hyperplasia 70
  infarct 65
  inflammation 69
  Ito cells 65
  liver weights 63
  necrosis 65–6
  neoplasia 70–2
  peliosis hepatis 69
  pigmentation 69
  steatosis 64
  spongiosis hepatis 69
lungs 112–120
  alveolar calcification 116
  alveolar hyperplasia 116–7
  bronchial associated lymph tissue (BALT)
    116
  hair embolism 117
  metastases 120
  mineralisation 117–8
  neoplasia 119–20
  oedema 113
  osseous metaplasia 119
  pleuritis 113
  pneumonia 113
  vascular calcification 117
luteinising hormone (LH) 15–16, 53, 146, 149,
    165, 184
luteoma 151–2
lymph nodes 129–37
  cystic degeneration 133
  histiocytosis 133–5
  hyperplasia 129–30
  lymphadenitis 131
  neoplasia 135–7
  sinus erythrocytosis 131–2

macrophages (alveolar) 114–6
malocclusion 17, 47
mammary glands 31–4
  hyperplasia 31–2
  inflammation 31
  neoplasia 33–4
medulloblastoma 225

melanocyte–stimulating hormone 184
meningiomas 225–7
meningitis 219
mesothelioma 154, 169
methylclofenapate 72, 75, 171
microlithiasis 88
microphthalmia 237
mineralisation
    adrenal 196
    brain 220
    heart 106
    intestines 59–60
    stomach 57
    testes 167
monocytic leukaemia 129
myeloid leukaemia 124–5
muscle 37–40
    atrophy 37–8
    inflammation 38
    necrosis 38
    neoplasia 39–40
myelofibrosis 124
myocarditis 98–101

nasal cavities 111
    rhinitis 111
    neoplasia 111
necrosis
    adrenal gland 196
    bone 41
    fat 25
    liver 65–6
    muscle 38
    parathyroid gland 208
    pituitary 185
    thymus 139
    vascular system 105
nematodes 59
nephroblastoma 90–1
nephrocalcinosis 88
nerve roots 231–2
nor-adrenaline 195, 201

obesity 6
oesophagus 55–6
    congenital anomalies 55
    neoplasia 56
    oesophagitis 55
oestrogen 34, 88, 123, 150, 156, 168, 185,
    189–90, 194, 220
oestrus cycle 146–8
oligodendroglioma 225–6
oncocytes 89
oral cavity 47–50
    neoplasia 48–50
organ weights 7–8
osseous metaplasia
    heart 101
    lung 119
    spleen 127

osteitis fibrosa cystica 84
otitis media 245
ovaries 145–154
    atresia 147
    atrophy 147
    ceroid 150
    corpora lutea 146–7
    cystic bursa 149
    cystic rete tubules 149
    food restriction effects 148–9
    functional changes 145–6
    inflammation 150
    interstitial cells 147
    neoplasia 150–4
    oestrus cycle 146–8
    ovarian weights 145
    pseudopregnancy 148
    stromal hyperplasia 150
oxytocin 184

pancreas (exocrine) 72–5
    altered foci 74
    atrophy 72–3
    hyperplasia 74
    neoplasia 74–5
parathyroid gland 208–9
    cysts 208
    hyperplasia 208
    inflammation 208
    necrosis 208
    neoplasia 209
Pasteurella pneumotropica 113
penis 178
peliosis hepatis 69
periodontitis 47
phaeochromocytoma 201
pigmentation
    adrenal 198
    kidney 89
    liver 69
    teeth 47
    thyroid 204
    uterus 155
pituitary gland 183–191
    castration cells 185
    cysts 185
    functional changes 184–5
    hyperplasia 185–6
    inflammation 185
    necrosis 185
    neoplasia 186–191
    pituitary weights 183–4
pneumonia 113
preputial gland 31
progesterone 15–16, 146, 148
prolactin 15–16, 34, 86, 168, 184–5, 189, 157
prostate gland 174–7
    atrophy 175–6
    corpora amylacea 176
    hyperplasia 176

inflammation 174
  neoplasia 176–7
  prostate weights 174
prostatitis 174
proteinurea 84–6
pseudopregnancy 148
pulmonary oedema 113–4
pyelonephritis 87–8
pyometra 155

quinoxaline 1,4 dioxide
  liver tumours 72
  kidney tumours 91
  nasal tumours 111

radiculoneuropathy 232
retinal atrophy 238–241
rhabdomyosarcoma 40

salivary glands 51–4
  cytomegaly 53
  functional changes 53
  inflammation 51
  neoplasia 54
salpingitis 156
sciatic nerve 232–4
spinal radiculoneuropathy 232–3
seminal vesicle 172–3
  atrophy 173
  inflammation 172
  neoplasia 173
sendai virus 3, 113
sertoli cell tubular hyperplasia 150
sertoli cell tumour 152
sertoliform tubular adenoma 152
sialoadenitis 51–2
siderofibrosis 127
sinus erythrocytosis 132
skin 23–28
  alopecia 24
  cysts 25
  inflammation 23
  neoplasia 26–8
  ringtail 24
spermatic granuloma 167, 171
spermatogenesis 164–5
spinal cord 230–2
  cysts 230
  degenerative myelinopathy 230–1
  neoplasia 232
spleen 126–9
  accessory spleen 126
  arteritis 128
  cysts 127
  extramedullary haematopoiesis 127
  lymphoid hyperplasia 127
  siderofibrosis 127
  spleen weights 126
spongiform encephalopathy 220–1
spongiosis hepatis 69

squamous metaplasia uterus 156
stomach 56–9
  ancanthosis 57
  erosions 56
  glandular dilatation 57
  hyperkeratosis 57
  mineralisation 57
  mucosal atrophy 57
  neoplasia 57–9
  ulceration 56–7

teratoma 151, 154, 169
testes 163–71
  arteritis 167
  atrophy 165–6
  cystic dilatation rete testis 167
  functional changes 164–5
  Leydig cell hyperplasia 168
  mineralisation 167
  neoplasia 168–71
  oedema 166
  spermatic granuloma 167
  testicular weights 163
testosterone 165, 168, 220
thrombosis 103
thymoma 139
thymus 137–42
  atrophy 138
  ectopic thymus 137
  necrosis 139
  neoplasia 139–42
  thymus weights 137
thyroglobulin 202
thyroid gland 201–8
  ectopic thymus 204
  C–cells 203–4
  cystic follicles 205
  functional changes 202–3
  hyperplasia 205–6
  hypertrophy 205
  inflammation 204
  neoplasia 206–8
  pigmentation 204
thyroid hormones 15–16, 37, 184, 185, 202–3
tongue 51
  inflammation 51
  neoplasia 51

urea 14
ureter 91
  hyperplasia 91
  inflammation 91
urinary bladder 91–3
  coagula 92
  cystitis 92
  neoplasia 93
urinogenital staining 17, 84
urolithiasis 88
uterus 154–8
  arteritis 155

Index

uterus (*continued*)
decidual reaction 156
endometrial hyperplasia 155
endometritis 155
functional changes 155
hydrometra 155
neoplasia 157–8
oestrus cycle 155
pigmentation 155
pyometra 155
squamous metaplasia 156
uterine weights 154

vagina 158–9
imperforate vagina 159
neoplasia 159
oestrus cycle 159
valvular degeneration 102

vascular system 105–7
arteritis 105–6
hair embolism 106
medial hypertrophy 106
mineralisation 106
neoplasia 106–7
thrombosis 106
vasculitis 172
vitamin A
liver 650
eye effects 238
stomach effects 57
vitamin B deficiency 233
vitamin E deficiency 233

Zymbal's glands 28–30
inflammation 28–30
neoplasia 30